Matthias Hermann Beringer

Design Studies towards a 4 MW 170 GHz Coaxial-Cavity Gyrotron

Karlsruher Forschungsberichte aus dem
Institut für Hochleistungsimpuls- und Mikrowellentechnik

Herausgeber: Prof. Dr. rer. nat. Dr. h.c. Manfred Thumm

Band 1

Design Studies towards a 4 MW 170 GHz Coaxial-Cavity Gyrotron

by
Matthias Hermann Beringer

Dissertation, Karlsruher Institut für Technologie
Fakultät für Elektrotechnik und Informationstechnik, 2010

Impressum

Karlsruher Institut für Technologie (KIT)
KIT Scientific Publishing
Straße am Forum 2
D-76131 Karlsruhe
www.ksp.kit.edu

KIT – Universität des Landes Baden-Württemberg und nationales
Forschungszentrum in der Helmholtz-Gemeinschaft

KIT Scientific Publishing 2011
Print on Demand

ISSN: 2192-2764
ISBN: 978-3-86644-663-2

Foreword of the Editor

Gyrotron oscillators (gyrotrons) are high-power millimeter-wave electron tubes mainly used for electron cyclotron resonance heating (ECRH) and plasma stabilization through localized non-inductive current drive in magnetically confined plasmas for energy generation by controlled thermonuclear fusion. Gyrotrons are electron cyclotron resonance masers operating in a longitudinal magnetic field, usually produced by a superconducting magnet. The current state-of-the-art for frequencies up to 170 GHz is 1 MW output power in continuous wave (CW) operation at an efficiency of approximately 50 % (employing a single-stage depressed collector for energy recovery). 2 MW gyrotrons with coaxial cavities for improved mode selection are under development for the worldwide fusion experiment "International Thermonuclear Experimental Reactor" (ITER) at Cadarache, France. In order to minimize the number of tubes and thus the number of superconducting magnets and to reduce the complexity of the whole heating system for future fusion reactors, gyrotrons with even higher output power are desirable. Such devices would reduce the space and costs requirements for the heating system significantly.

In this work of Dr.-Ing. Matthias H. Beringer the feasibility of a 4 MW 170 GHz coaxial-cavity gyrotron for CW operation has been demonstrated and detailed physical designs of the major components have been elaborated. Since synthetic diamond output windows allow the transmission of up to 2 MW, CW, the design of the 4 MW tube contains two output windows and two 2 MW output millimeter-wave beams. In a first step, one possible very high order cavity operating mode was selected, followed by designs of the major gyrotron components: electron gun, coaxial cavity, two-beam quasi-optical output coupler and collector. Several thermo-mechanical studies were performed to identify possible problems regarding cooling techniques and CW operation. These calculations include the deformation of the cavity and the longitudinal impedance corrugation on the coaxial cavity insert as well as the surface perturbations of the quasi-optical waveguide antenna launcher of the output coupler.

With this PhD thesis Dr.-Ing. Matthias H. Beringer provides the worldwide first detailed feasibility study of a 4 MW, CW fusion gyrotron which will be a solid basis for the final complete engineering design for such a tube. We wish Dr. Beringer much of further success in his professional career and we are sure that his future activities will benefit from his excellent abilities.

Design Studies towards a 4 MW 170 GHz Coaxial-Cavity Gyrotron

Zur Erlangung des akademischen Grades eines

DOKTOR-INGENIEURS

von der Fakultät für
Elektrotechnik und Informationstechnik
am Karlsruher Institut für Technologie (KIT)

genehmigte

DISSERTATION

von

Dipl.-Ing. Matthias Hermann Beringer
aus Poppenlauer

Tag der mündlichen Prüfung: 15.12.2010

Hauptreferent: Prof. Dr. rer. nat. Dr. h.c. Manfred Thumm
Korreferent: Prof. Dr.-Ing. Mathias Noe

Kurzfassung

In der vorliegenden Arbeit wird die Machbarkeit eines 4 MW 170 GHz Gyrotrons mit koaxialem Resonator gezeigt und entsprechende physikalische Designs der wichtigsten Gyrotronkomponenten für lange Pulse erarbeitet. Zukünftige thermonukleare Fusionskraftwerke benötigen zuverlässige und leistungsfähige Millimeterwellenquellen für die Heizung und Stabilisierung der Kernfusionsplasmen. Die Entwicklung von Gyrotrons mit hoher Ausgangsleistung reduziert die hierzu nötige Anzahl der Quellen und die der entsprechend erforderlichen supraleitenden Magnete.
Zunächst wird eine passende Arbeitsmode ermittelt, welche eine effiziente Elektron-Zyklotron-Wechselwirkung und gleichzeitig niedrige ohmsche Wandverluste im Resonator garantiert. Die gewählte Mode vereinfacht gezielt das Design des zweistrahligen quasi-optischen Wellentypwandlers und begünstigt zusätzlichen Mehrfrequenzbetrieb. Die optimierte Geometrie des koaxialen Resonators ermöglicht eine stabile Einmodenschwingung bei gewünschter Ausgangsfrequenz und -leistung. Elektronenkanonen (sog. magnetron injection guns) in Ausführungen als Diode und Triode werden erarbeitet, um den für die Wechselwirkung mit hohem Wirkungsgrad nötigen helikalen Elektronenstrahl bereit zu stellen. Speziell das Layout der Triode eröffnet eine neuartige Möglichkeit deren Größe zu reduzieren und die Installation in einem preisgünstigeren supraleitenden Magneten mit kleinem Bohrloch zu ermöglichen. Eine erweiterte Methode wird aufgezeigt, um die Rundhohlleiterantenne zu optimieren, welche die generierte Resonatormode hoher Ordnung in zwei Ausgangsstrahlen mit Gaußschem Profil wandelt. Die Oberfläche des Kopplers wird analysiert und der Einfluss systematischer Techniken zur Glättung der Wandstruktur wird charakterisiert. Zur Absorption des Elektronenstahls nach der Wechselwirkung und zusätzlicher Energierückgewinnung werden zwei Entwürfe für Kollektoren, mit reinem und dispersionsverstärktem Kupfer als Oberflächenmaterial, aufge-

zeigt und entsprechende Verbesserungsansätze diskutiert.
Zusätzlich zu den Designs der wichtigsten Gyrotronkomponenten werden mehrere thermo-mechanische Studien durchgeführt um kritische Bereiche der Hitzeentwicklung zu identifizieren und entsprechende Kühltechniken zu verifizieren. Die Betrachtungen beinhalten die thermische Deformation des Resonators, der Impedanzrillung am koaxialen Innenleiter und der Perturbationen des quasi-optischen Wellentypwandlers.

Abstract

In this work the feasibility of a 4 MW 170 GHz coaxial-cavity gyrotron is demonstrated and proper physical designs of the major gyrotron components suitable for long-pulse operation are elaborated. Future thermonuclear fusion power plants demand highly reliable and efficient millimeter wave sources for heating and stabilization of the fusion plasma. The realization of gyrotrons with high unit power will reduce the overall number of sources and superconducting magnets required.
As a first step, a mode selection procedure is presented in order to identify operating parameters, which guarantee efficient electron-cyclotron-interaction and low ohmic wall loading inside the cavity. The selected mode is determined to simplify the design of a two-beam output coupler and possible multi-frequency operation. The optimized coaxial cavity is suitable to provide the desired output power in stable single-mode operation. Diode- and triode-type magnetron injection guns are designed in order to deliver the high quality helical electron beam necessary for interaction. Especially the triode-type electron gun shows a novel possibility for the reduction of the space requirements which allows to utilize a smaller warm bore hole of the superconducting magnet and thus reduce its costs. A new possibility to design the quasi-optical output coupler, which transforms the generated high-order volume mode into two Gaussian beams, suitable for the two output windows, is developed. The launcher's contour surface structure is analyzed and the influences of systematic smoothening techniques on the shape of the wall's perturbation are studied. Two depressed collector layouts, using pure and dispersion strengthened copper, are elaborated to absorb the spent electron beam.
In addition to the component designs, several thermo-mechanical studies are performed in order to identify critical domains of thermal treatment and the corresponding cooling techniques required. The considerations include

the deformation of the cavity, the impedance corrugation on the coaxial insert and the launcher surface perturbations.

Contents

List of used Symbols and Variables

Common indices and operators

$\perp$	perpendicular or at cutoff
$\parallel$	parallel or towards longitudinal direction
E, C	at the emitter or at the electron gun's cathode
i	coaxial insert
L	at the launcher
m, p	mode indices
n	axial mode index
R	at the cavity (resonator)
S	on surface S
$*$	conjugated conversion of a complex function
∇	Nabla operator with $\nabla = (\partial/\partial x_1)\,\vec{e}_1 + \cdots + (\partial/\partial x_n)\,\vec{e}_n$
Δ	Laplace operator with $\Delta = \nabla \cdot \nabla$

List of variables

A, B	arbitrary constants or field amplitudes
$\vec{A}$	magnetic vector potential with $\vec{B} = \nabla \times \vec{A}$
A_φ	azimuthal component of magnetic vector potential
A_G and A_0	material constants
b	magnetic compression ratio with $b = B_R/B_E$ or basis functions
$\vec{B}$	magnetic field vector (magnetic flux density) with amplitude B
B_E	magnetic field amplitude at emitter
B_R	magnetic field amplitude at cavity
c	sequence of transformation coefficients

c_0	speed of light in vacuum $c_0 = 2.9979 \cdot 10^8\,\mathrm{m/s}$
C	ratio of the cavity's radii with $C = R_\mathrm{R}/R_\mathrm{i}$
d_col	thickness of collector wall
d_m	depth of cavity outer wall corrugation
e	elementary electron charge $e = 1.6022 \cdot 10^{-19}\,\mathrm{C}$
$\vec{e}$	eigenvector
E_φ and E_r	transverse electric field components
E_t	tangential electric field component
$\vec{E}$	electric field vector
$\vec{E}_\mathrm{E}$	electric field vector at emitter
E^i and E^s	incident and outgoing (scattered) wave
f_0	gyrotron output frequency
f_c	relativistic cyclotron frequency $f_\mathrm{c} = \Omega_\mathrm{c}/2\pi$
f_sw	collector sweeping frequency
Δf	frequency shift
$f(z)$	complex longitudinal field profile
$\vec{F}_\mathrm{L}$	Lorentz force
g	Green's function
h	Planck's constant $h = 6.626 \cdot 10^{-34}\,\mathrm{Js}$
h_c	heat exchange coefficient
H	filter function
$\vec{H}$	magnetic field vector (magnetic field strength) with $\vec{H} = \vec{B}/\mu_0$
H_m	m-th order Hankel function
i	iterative number
I	current or constant number
I_b	beam current
I_lim	limiting current for $\beta_\| \to 0$
$I_\mathrm{b,n}$	beam current of n-th particle
$I_\mathrm{s,n}$	sweeping current of n-th collector coil
$\vec{I}_\mathrm{S}$	surface current
j	imaginary unit
$\vec{j}$	complex current density
j_E	current density at emitter

J_m	m-th order Bessel function
k and k_0	iterative number or wave number with $k = 2\pi/\lambda = 2\pi \cdot f/c_0$
$k_\perp = (k_\mathrm{r})$ and $k_\parallel$	perpendicular and parallel wave number with $k_0^2 = k_\parallel^2 + k_\perp^2$
k_B	Boltzmann constant with $k_\mathrm{B} = 1.3807 \cdot 10^{-23}\,\mathrm{m^2kg/s^2K}$
k_sr	factor representing the surface roughness
K	constant factor
$\vec{dl}$	infinitesimal vector along Σ
Δl	length of deformation
L_B	Brillouin length
L_E	axial emitter length
L_L	length of launcher
L_d and L_u	length of down- and uptaper section
L_R	length of cylindrical cavity section
$L_\mathrm{S,d}$ and $L_\mathrm{S,u}$	length of parabolic smoothing towards up- or downtaper
m	azimuthal mode index or filter order
m_e	electron rest mass $m_\mathrm{e} = 9.11 \cdot 10^{-31}\,\mathrm{kg}$
M	arbitrary number
$\vec{n}$	unit vector
n	axial mode index
N	number of turns or arbitrary number
N_m	m-th order Neumann function
p	radial mode index or complex electron momentum
$p_\perp$	perpendicular complex electron momentum
P	power or separated function
P_elec	generated RF-power in the cavity
$P_\perp$	perpendicular electron power
P_out	output power
q	particle charge
Q	quality factor, reactive power or separated function
Q_diff	diffractive quality factor

Q_{ohm}	ohmic quality factor
Q_{tot}	total quality factor $1/Q_{\mathrm{tot}} = 1/Q_{\mathrm{diff}} + 1/Q_{\mathrm{ohm}}$
r	radius or ratio between collector wobbling and sweeping frequency
r_{L}	Larmor radius
r_{s} and r_{p}	attenuation in stop- and passband
R_{b}	average beam radius
$R_{\mathrm{b,R}}$	average beam radius at cavity
$\Delta R_{\mathrm{b,rms}}$	root mean square spread of R_{b}
R_{c}	caustic radius
R_{E}	emitter mean radius
ΔR_{G}	widening of the beam's guiding centers
R_{i}	radius of coaxial insert
R_{L}	launcher radius
$R_{\mathrm{L,0}}$	untapered launcher radius
ΔR_{L}	launcher surface perturbations
R_{R}	radius of cavity (resonator)
ΔR_{R}	thermal cavity expansion
Re	Reynold's number
$d\vec{s}$	infinitesimal surface vector
s	number of cyclotron harmonic
S	surface area or apparent power
t	time variable or transformation matrix
T	temperature or Fourier transform of t
u	field on unperturbated launcher surface
u_{c}	field on perturbated launcher surface
$\vec{u}$	electron momentum
u_{φ}	angular electron momentum
$u_{\parallel}$	parallel electron momentum
U	vector potential
$\vec{v}$	velocity vector with amplitude v
$\vec{v}_{\perp}$ and $\vec{v}_{\parallel}$	parallel and perpendicular velocity components
V	voltage
V_{acc}	accelerating voltage

V_b	beam energy in terms of eV ($V_\mathrm{acc} - \Delta V$)
V_depr	depression voltage
V_mod	modulation voltage for triode-type electron gun
ΔV	voltage drop due to voltage depression with $\Delta V = |V_\mathrm{acc} - V_\mathrm{b}|$
W	energy or working function
W_kin	kinetic energy
x_n and y_n	single discrete values
X and Y	random variable
z_L	axial distance during one Larmor period
Z	separated function
Z_0	free space impedance with $Z_0 = \mu_0 c_0 = \sqrt{\mu_0/\varepsilon_0} = 1/(\varepsilon_0 c_0) \approx 376.73\,\Omega$

α	velocity ratio (pitch factor) with $\alpha = v_\perp/v_\parallel = \beta_\perp/\beta_\parallel$
β	normalized velocity with $\beta = v/c_0$ and $\beta^2 = \beta_\perp^2 + \beta_\parallel^2$
$\beta_\perp$ and $\beta_\parallel$	normalized perpendicular and parallel velocity components
$\Delta\beta$	spread of normalized velocity
$\Delta\beta_{\perp,\mathrm{rms}}$ and $\Delta\beta_{\parallel,\mathrm{rms}}$	root mean square spread of normalized velocity component
$\Delta\beta_{\perp,\mathrm{max}}$ and $\Delta\beta_{\parallel,\mathrm{max}}$	maximum spread of normalized velocity component
γ	relativistic factor (Lorentz factor)
Γ	complex reflection coefficient
δ	skin depth
Δ	normalized detuning
ε_0	vacuum permittivity with $\varepsilon_0 = 8.85 \cdot 10^{-12}\,\mathrm{F/m}$
ζ	normalized axial position
η	efficiency or vector correlation coefficient

η_{elec}	efficiency of energy removal from electrons with $\eta_{\mathrm{elec}} = \alpha^2/(1+\alpha^2) \cdot \eta_\perp$
$\eta_\perp$	perpendicular efficiency with $\eta_\perp = P_{\mathrm{elec}}/P_\perp$ and downtapers
θ_{R}, θ_{d} and θ_{u}	angles for resonator or linear up-
Θ	probability function
λ	wavelength
λ_{R}	material constant
Λ	slow variable phase
μ_0	permeability of free space with $\mu_0 = 4\pi \cdot 10^{-4}\,\mathrm{H/m}$
ν	arbitrary factor for coil current
ξ	evolution factor or phase offset
π	ratio of a circle's circumference to its diameter $\pi \approx 3.14159265$
ϖ	arbitrary factor for coil current
ρ	space charge or wall loading
σ and σ^2	material conductivity or standard deviation
Σ	contour surrounding surface S
τ	launcher slope or scaling factor
φ_{B}	angle of magnetic field vector relative to z-axis
φ_{C}	slant angle of cathode
φ_{EB}	angle between el. $\vec{E}_{\mathrm{E}}$ and magn. field $\vec{B}_{\mathrm{E}}$ at emitter
$\varphi_{\mathrm{l,s}}$	solid angle at impedance corrugation of coaxial insert
ϕ	phase of sweeping current
Φ_{m}	magnetic flux
$\chi_{\mathrm{m,p}}$	eigenvalue of mode
ψ	Brillouin angle
ω_{RF}	frequency of RF-oscillation
ω_{n}, ω_{s} and ω_{p}	sampling rate, stop- and passband frequency
Ω_0	nonrelativistic cyclotron frequency with $\Omega_0 = eB/m_{\mathrm{e}}$
Ω_{c}	relativistic cyclotron frequency with $\Omega_{\mathrm{c}} = 2\pi f_{\mathrm{c}} = \Omega_0/\gamma$
Ω_{f}	frequency of fast changing phase

1. Introduction – Gyrotrons for thermonuclear fusion applications

Within the years 2005 to 2015 the UN predicts a growth rate of 79 million p.a. for the world population. As a consequence, depending on the utilized scenarios, the world population will increase from nearly 7 billion people at present (2010) to as many as 10 billion people by 2050 [Nat04]. This rapid growth and the finite reservoir of fossil fuels require new methods of environmentally friendly energy generation. One possible future candidate is the energy gain due to thermonuclear fusion of magnetically confined plasmas, in which two atoms with light nuclei are fused and the mass decrease is converted to energy in agreement with $E = \Delta m c_0^2$. For technical application the fusion of a deuteron with a triton creating an alpha-particle, freeing a neutron, and releasing 17.6 MeV of energy, is of interest.

$$d^+ + t^+ \rightarrow \alpha^{2+}(3.5\,\text{MeV}) + n(14.1\,\text{MeV}) \tag{1.1}$$

A substantial energy barrier (Coulomb barrier) of electrostatic forces must be overcome before the residual strong nuclear force takes over and fusion can occur. This can be achieved by heating the educts for the reaction to a temperature of several billion Kelvin, resulting in a plasma, in which the electrons are separated from their nuclei. The hot fusion plasma can be confined in a magnetic configuration, such as the toroidal geometries of so-called tokamaks and stellarators.

To achieve the required high temperatures, nuclear plasmas in fusion reactors can be efficiently heated by microwaves. The electron-cyclotron-resonance-heating (ECRH) heats initially the electrons, which transfer their energy via Coulomb collisions to the ions in the plasma. Microwave heating has several advantages in comparison to other heating techniques. Inter alia, ECRH requires small antennas towards the vacuum chamber, the coupling to the plasma can be up to 100% and the area of absorption within the plasma can be localized very well. Absorption of the microwave energy is most efficient when the frequency is close to the cyclotron frequency of the electrons in the magnetically confined plasma. To get reasonable plasma radii within the magnetic field confinement, cyclotron frequencies of approximately 100 GHz to 200 GHz and, for future reactor designs, power levels of several 10 MW are required. Today, gyrotrons (gyro-oscillators or gyromonotrons) are the only feasible coherent high-power microwave sources which can operate efficiently and reliably within this power and frequency region. Unlike classical "slow-wave" microwave tubes (magnetron, klystron etc.), the gyrotron operates in a high-order mode and thus the dimensions of the resonant cavity are not limited to the scale of the emitted radiation's wavelength. The output power can therefore greatly exceed that of conventional microwave tubes.

Current gyrotron development has demonstrated continuous wave (CW) output power per unit up to $1 - 2$ MW [Thu10]. To minimize the number of heating units required and to reduce the complexity of the whole heating system for future fusion reactors, gyrotrons with even higher output power are desirable. Such devices will reduce the costs and space requirements for the heating system significantly.

In this work the feasibility of a 4 MW 170 GHz gyrotron is studied and detailed physical designs of the major components are elaborated. In particular, a possible reduction of the electron gun size, a novel layout for a two-beam quasi-optical output coupler and the feasibility of a 4 MW collector have been obtained. Several thermo-mechanical investigations are performed to identify possible challenges of long-pulse operation effects in order to deliver reliable and efficient sources, as they gain increasing importance for the successful realization of future fusion reactors.

1.1. Current state of research and the 4 MW power level

Gyrotron oscillators are high-power microwave sources mainly used for electron-cyclotron-resonance-heating and plasma stabilization through localized current drive in magnetically confined plasmas for controlled thermonuclear fusion experiments. At the Institute for Pulsed Power and Microwave Technology (IHM) of the Karlsruhe Institute of Technology (KIT), high-power gyrotrons are under development in collaboration with several partners [Thu10]: The world record parameters of the European 140 GHz conventional-cavity gyrotron are 0.92 MW at 1800 s pulse duration and 44 % efficiency, employing a single-stage depressed collector for energy recovery. The short-pulse pre-prototype tube of the European 2 MW 170 GHz coaxial-cavity gyrotron for ITER achieved a power of 2.2 MW at 30 % efficiency, without a depressed collector. The energy record is held by the JAEA-Toshiba 170 GHz gyrotron with 0.8 MW for 3600 s [SKK$^+$09]. State-of-the-art high-power tubes are also being developed at the Institute of Applied Physics in Nizhny-Novgorod, Russia [DLM$^+$08]. After its discovery [Twi58][Sch59][Gap59], the application of the electron-cyclotron-interaction was mainly pioneered in the former Soviet Union in order to build powerful electronic vacuum tubes.

Future fusion facilities and reactors require highly efficient and reliable heating systems. Increased unit output power would reduce the overall number of gyrotron devices required. This might lead to considerable cost savings because fewer superconducting magnets are necessary and both, the complexity of a possible ECRH installation and the space requirements, are decreased.

Investigations on the feasibility of 4 MW gyrotrons should identify the challenges and restrictions at even higher power levels for the various components, but keep in mind the difficulties and challenges in the present developments.

1.2. Content and structure of this work

After selection of a suitable operating mode, complete designs for the major components of a 4 MW 170 GHz coaxial-cavity gyrotron for continuous wave (CW) operation are developed and thermo-mechanical studies concerning long-pulse operating effects are performed.

At the beginning the main components and the operational principle of gyro-oscillators are introduced. In the second chapter a mode selection procedure is performed, identifying operating parameters which guarantee efficient electron-cyclotron-interaction and low ohmic wall loading inside the cavity. In addition the desired mode should be suitable to simplify the design of a two-beam output coupler (launcher) and possible multi-frequency operation. In a next step, the employed codes to describe the interaction are verified and critical numerical parameters are elaborated. The optimization of a coaxial cavity including nonlinear input and output tapers is described.

For the design of the electron gun as either a diode- or triode-type setup in chapter three, automated optimization algorithms are introduced to provide geometries and magnetic setups which deliver a high-quality helical electron beam. The numerous electrical and geometrical gun parameters are studied and statements concerning tolerances and reliability are worked out. At the necessary high electric field values of the 4 MW operation point, triode-type guns show a novel possibility for the reduction of their space requirements, which allows one to utilize a smaller warm bore hole of the superconducting magnet and thus reduce its costs.

In chapter five, a new possibility to design the quasi-optical output coupler, which transforms the generated high-order volume mode into two beams, suitable for the two output windows, is developed. The launcher's contour surface structure is analyzed and the influences of systematic changes in the shape of the wall perturbations are studied. Several smoothing methods are introduced using two-dimensional low-pass filter techniques.

Chapter six describes the feasibility and the design process of two collector layouts for the 4 MW operation using pure and dispersion strengthened copper as surface material. The studies include the optimization of the surface shape as well as possible longitudinal sweeping systems.

In addition to the complete designs of the major gyrotron components, several thermo-mechanical studies are performed. The results are obtained using commercial finite-element codes in combination with newly developed scripts for the data exchange with corresponding KIT in-house codes. As a first step in chapter seven, the expansion of the designed cavity and the corresponding shift in the output frequency are studied. During operation the coaxial insert and its impedance corrugation are heated and deformed which might lead to reduced performance if the suppression of unwanted competing modes is less efficient. In a last step, simulations regarding the heat treatment of the fine perturbations on the launcher surface are carried out. The reduced efficiency of mode conversion for a deformed launcher surface is analyzed and parameter studies are performed. Different scenarios for the alignment of the launcher's cooling channels are evaluated.

1.3. The gyro-oscillator – main components and operation principle

The gyrotron oscillator is a vacuum electron tube working as a electron-cyclotron maser (microwave amplification by stimulated emission of radiation). Electronic vacuum devices are still unmatched in their performance as coherent high-power sources and amplifiers for electromagnetic radiation in the domain of milli- and submillimeter waves. Gyrotron devices can produce radiation with output powers ranging from tens of watts to several megawatts. The covered frequency spectrum spans from the lower GHz-towards the THz-region.

The schematic of a high-power gyrotron with a coaxial cavity and a two-beam output is shown in Fig. 1.1. The use of a coaxial insert has considerable advantages for high-power gyrotrons in comparison to conventional hollow waveguide structures as described in Sec. 1.3.3. In principle, electrons are extracted from a ring-shaped emitter on the cathode and accelerated towards the interaction cavity by an electric field generated due to the voltage drop across cathode and anode. The electron trajectories follow the magnetic field lines provided by the surrounding magnet system. In the cavity the electron-cyclotron-interaction takes place and the RF-power is generated. Within the quasi-optical output system the output beam(s) are formed and separated from the electron trajectories. The spent electron beam is absorbed in the collector and the microwaves exit the tube through one or more output windows.

Starting from the electron-cyclotron-interaction, in the following sections the main components of a gyrotron (electron gun, cavity, output coupler and collector) are introduced.

1.3.1. Electron-cyclotron-interaction

Starting from the emitter, each electron is forced onto a helical orbit by the magnetic field $\vec{B}$ produced by a system of superconducting magnets surrounding the tube (see Fig. 1.3). Their rotational frequency f_c (relativistic cyclotron frequency) and Larmor radius r_L can be calculated by equating the Lorentz- with the centripetal-force.

$$\Omega_c = 2\pi f_c = \frac{eB}{m_e \gamma} \approx 2\pi \cdot \frac{28\,\text{GHz} \cdot B/T}{\gamma} \tag{1.2}$$

$$r_L = \frac{v_\perp}{\Omega_c} = \frac{m_e \gamma v_\perp}{eB} = \frac{v_\perp}{2\pi f_c} \quad \text{with} \quad \vec{v} = \vec{v}_\perp + \vec{v}_\parallel \tag{1.3}$$

The electrons are described by their mass m_e and charge $-e$, parallel $v_\parallel$ and perpendicular $v_\perp$ velocity components. Due to the high accelerating voltage V_{acc} (in the $100\,\text{kV}$-range for high-power gyrotrons), the electrons reach a weakly relativistic velocity v with the Lorentz factor γ given as:

$$\gamma = \frac{1}{\sqrt{1 - (v/c_0)^2}} = 1 + \frac{W_{\text{kin}}}{m_e c_0^2} \approx 1 + \frac{W_{\text{kin}}}{511\,\text{keV}} \tag{1.4}$$

W_{kin} denotes the kinetic energy and c_0 the speed of light in vacuum.

With an additional RF-field at the frequency ω_{RF}, which is nearly synchronous to the rotation of the particles, some electrons lose energy to the RF-field and some gain energy, depending on their relative phase position. While at nonrelativistic energies, the net energy exchange between

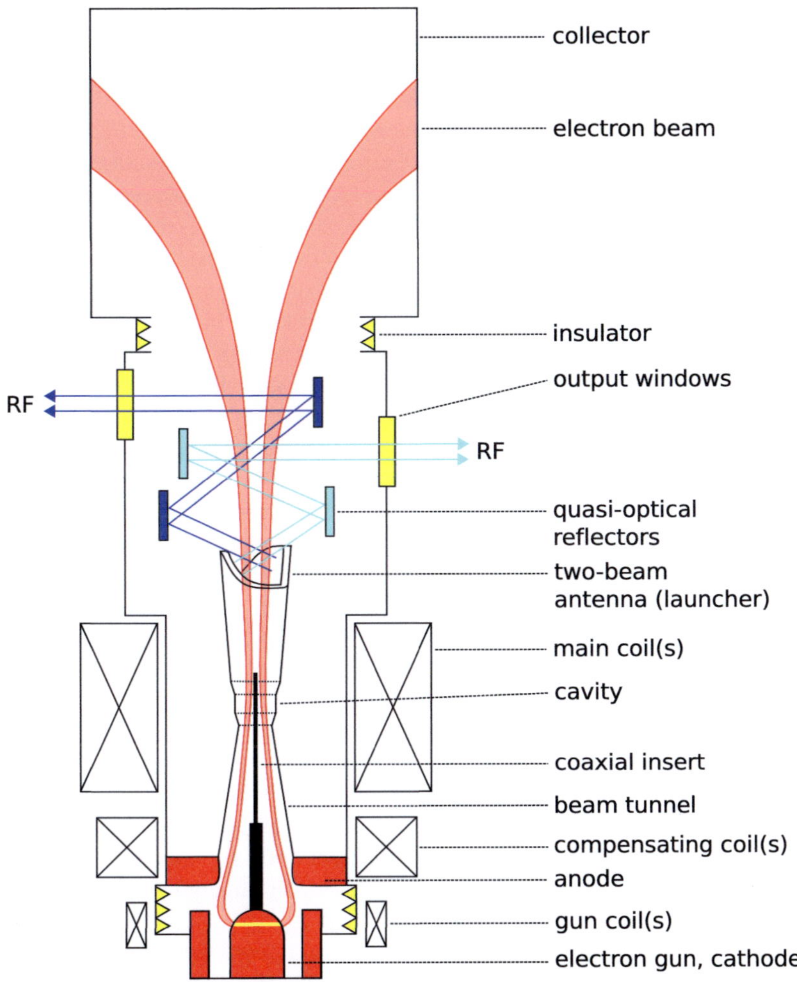

Fig. 1.1.: Sketch of a coaxial-cavity gyrotron with two-beam output and magnet system

the whole electron beam and RF-field averages to zero, the changing relativistic factor γ of the individual electrons can cause a non-zero net energy transfer and thus permits the generation and enhancement of the RF-field. The electron-cyclotron-interaction, as described in the following, is therefore a relativistic effect. As a consequence of the changing electron energy and the corresponding cyclotron frequency change, the frequency difference $|\omega_{RF} - \Omega_c|$ between RF-field and electron movement changes and thus the relative phase difference for every particle varies over time.

Assuming that initially ω_{RF} is slightly higher than Ω_c, the behavior of electrons initially losing energy differs from that of electrons initially gaining energy. Electrons in an accelerating phase position increase γ, thus decrease their Ω_c and increase the difference with respect to the initial higher ω_{RF}. Over time, they accumulate an increasing phase lead with which they tend to leave this undesired phase position more quickly. In contrast, as indicated in Fig. 1.2, decelerated particles increase their Ω_c and decrease the difference with respect to ω_{RF}, which means they tend to remain longer in their beneficial initial phase position relative to the RF-field.

In summary, the particles accumulate in a particular relative phase position where they transfer energy to the RF-field. This focusing towards one phase is called "bunching". As mentioned before, to enable a beneficial energy exchange, a priori the oscillation frequency ω_{RF} has to be slightly higher than the angular velocity Ω_c of the electrons.

$$\omega_{RF} \gtrsim \Omega_c \qquad (1.5)$$

If the electrons stay too long under the influence of the RF-field, they leave the beneficial phase position again. Consequently the electrons' energy transfer decreases, and this process is called "overbunching".

In addition to this simplified description of the electron-cyclotron-interaction, the motion $v_{\parallel}$ of the electrons relative to the RF-field has to be considered. In response to $v_{\parallel}$, the RF-frequency ω_{RF} which acts on the electrons is shifted by a Doppler term. To describe the Doppler term the wave number k is introduced, which gives the number of wavelengths per unit distance.

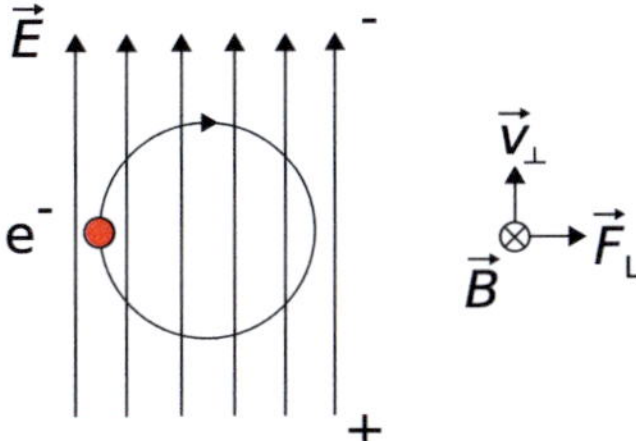

Fig. 1.2.: Electron with cyclotron frequency Ω_c loses energy towards RF-field with ω_RF

$$k_0 = \frac{2\pi}{\lambda} = \frac{2\pi f}{c_0} \tag{1.6}$$

The total wave number k_0 can be split into two portions for the wave propagation in the axial $k_\parallel$ and perpendicular $k_\perp$ directions.

$$k_0^2 = k_\parallel^2 + k_\perp^2 \tag{1.7}$$

The phase of the field $(\omega_\mathrm{RF} t - k_\parallel z)$ at the position of the electron, which is also moving towards positive z-positions, should be synchronous to the rotational particle movement $\Omega_\mathrm{c} t$. As a consequence, the corresponding phase difference should nearly not vary over time.

$$\frac{d}{dt}\left(\omega_\mathrm{RF} t - k_\parallel v_\parallel t - \Omega_\mathrm{c} t\right) \gtrsim 0 \tag{1.8}$$

Considering the Doppler shift, the resonance condition is derived as:

$$\omega_\mathrm{RF} - k_\parallel v_\parallel \gtrsim \Omega_\mathrm{c} \tag{1.9}$$

Oscillations are also possible at higher cyclotron harmonics $s \cdot \Omega_\mathrm{c}$, in which the RF-field is inhomogeneous along the electron rotation. As a consequence, the general oscillation condition is:

$$\omega_\mathrm{RF} - k_\parallel v_\parallel \gtrsim s \cdot \Omega_\mathrm{c} \tag{1.10}$$

Usually the gyrotron is operated close to cutoff in order to generate highest possible amplitudes for the perpendicular electric field components of the desired TE-mode. As a result, $k_\parallel \to 0$ close to cutoff, and the influence of the Doppler shift becomes small.

1.3.2. Electron gun and beam tunnel

The preceding discussion shows that the electron-cyclotron-interaction extracts energy for the RF-field only out of the perpendicular component of the electron motion. Thus electrons with a high perpendicular velocity component $v_\perp$ are necessary for efficient electron-cyclotron-interaction. In addition, their longitudinal velocity $v_\parallel$ has to be sufficiently high to enable the particles to reach the interaction region and leave it again after energy transfer. In high-power gyrotrons usually *magnetron injection guns* (MIGs) are used to provide the necessary hollow electron beam. An annular beam configuration allows stable interaction with high-order waveguide modes in cylindrical geometries [Edg99]. Fig. 1.3 shows the cross-section of a typical triode-type MIG for a coaxial gyrotron and the electron beam with the radial extension of the guiding centers. The setup is surrounded by a superconducting magnet system: The main coils provide the necessary high axial field in the cavity. With a compensating coil the field produced by the main coils is suppressed in the emitter region. The gun coils provide the possibility for sensitive tuning the desired magnetic field at the starting points of the electron trajectories.

The electrons are accelerated starting from a heated emitter ring towards the beam tunnel and cavity. The dispersion cathode typically consists of porous tungsten filled with special oxides to provide a low electron affinity. According to the characteristics of the magnetic field, the emitter forms a hollow beam, whose radius R_b is substantially larger than the radius of the electrons' rotation in their helical orbits. The MIGs used in gyrotrons are usually operated under temperature-limited (TL) condition of the emitter [Edg99]. A non-zero electric field $\vec{E}_E$ exists at the cathode surface and accelerates all emitted electrons towards the anode (or modulation anode for the triode). The beam current depends sensitively on the conditions at the emitting surface and can be adjusted by the temperature of the emitter.

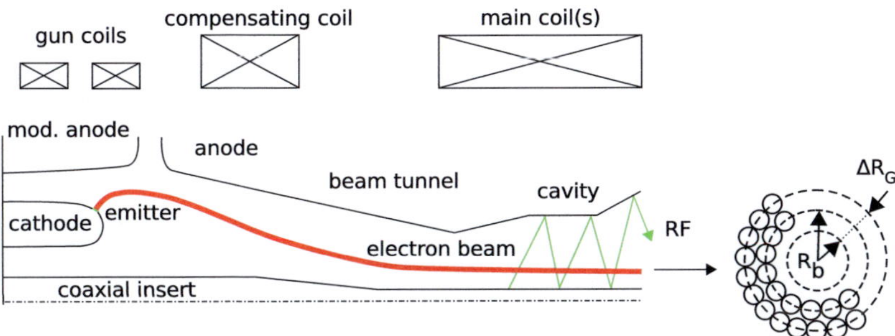

Fig. 1.3.: Axially symmetric cross-section of a triode-type magnetron injection gun

An important consideration for long lifetime of the emitter material is the maximum allowable current density. In modern emitters, the technically attainable limit is about 5 to $10\,\mathrm{A/cm^2}$. The electrons' perpendicular (rotational) $v_\perp$ and axial (parallel) $v_\parallel$ velocity components respectively are usually given normalized to the speed of light in free space c_0.

$$\beta_\perp = \frac{v_\perp}{c_0} \quad \text{and} \quad \beta_\parallel = \frac{v_\parallel}{c_0} \tag{1.11}$$

The total normalized electron velocity is $\beta = \sqrt{\beta_\perp^2 + \beta_\parallel^2}$ as indicated in Fig. 1.4(a)). The related velocity ratio (also called "pitch factor") α is one key parameter for the characterization of the electron beam.

$$\alpha = \frac{\beta_\perp}{\beta_\parallel} \tag{1.12}$$

Due to the increase of the axis-symmetric static magnetic flux density from $|\vec{B}_{\mathrm{E}}|$ at the emitter towards $|\vec{B}_{\mathrm{R}}|$ at the interaction cavity, the electron beam radius R_{b} is compressed and energy is transferred towards the perpendicular velocity component, which means α is increased according to the increasing cyclotron frequency. Typically high-power gyrotrons are operated with electron beams having a velocity ratio α between 1.2 and 1.5.

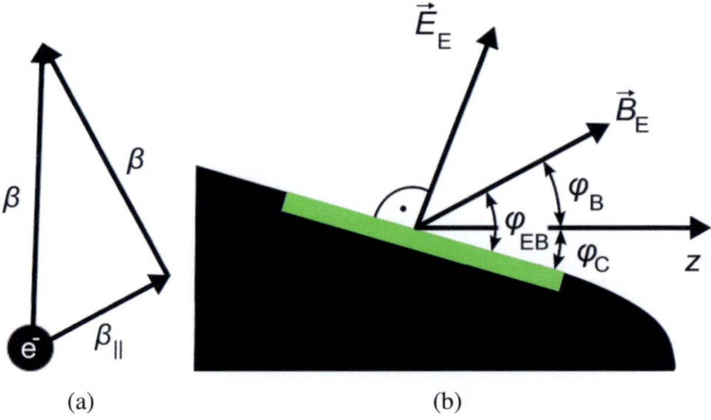

Fig. 1.4.: Velocity components (a) at arbitrary position and field vectors (b) at the emitter (green)

The compression ratio b of the beam radius from the emitter radius R_E towards its desired average value at the cavity center $R_{b,R}$ is given as:

$$b = \frac{B_R}{B_E} = \left(\frac{R_E}{R_{b,R}} \right)^2 \tag{1.13}$$

$\eta_\perp$ is the efficiency of the energy transfer from the perpendicular electron power $P_\perp$ of the gyrating electrons to the generated RF-power in the cavity P_{elec}.

$$\eta_\perp = \frac{P_{elec}}{P_\perp} \tag{1.14}$$

$\eta_\perp$ has a theoretical upper limit of $\approx 72\%$ [KDST85], based on the *theory of normalized variables* as described in Sec. 2.3. The efficiency of the total energy removal from the electrons is given by η_{elec}.

$$\eta_{elec} \approx \frac{\alpha^2}{1 + \alpha^2} \cdot \eta_\perp \tag{1.15}$$

The properties of the electron beam are highly sensitive to the magnetic and electric field in the vicinity of the emitter. Under the assumptions of the *adiabatic theory*, as shown in Ch. 3 or more detailed in [Edg99], the transverse velocity of electrons at the interaction region depends on the operating parameters in first approximation as:

$$\beta_\perp \approx \frac{1}{\gamma \cdot c_0} \sqrt{b} \cdot \frac{E_E \cos(\varphi_{EB})}{B_E} = \frac{1}{\gamma \cdot c_0} \sqrt{\frac{B_R}{B_E^3}} \cdot E_E \cos(\varphi_{EB}) \qquad (1.16)$$

The related parameters can be found in Fig. 1.4(b). φ_C is the slant angle of the cathode and φ_B is the angle of the magnetic field vector $\vec{B}_E$ related to the z-axis. The sum φ_{EB} is the angle between the direction of the magnetic field and the emitter surface. As mentioned before, with the gun coils of the magnetic setup, which are close to the emitter, φ_{EB} can be smoothly tuned without modifying the magnitude of the magnetic field and consequently without change of the compression ratio b.

The electrons extracted from the emitter gain kinetic energy from the electric field produced by the accelerating voltage V_{acc} as they move downstream to the resonator. Due to the space charge of the electron beam, a negative potential is created which screens the electrons partially from the accelerating voltage. This effect is known as *voltage depression*. For coaxial MIGs, the voltage drop ΔV reduces the accelerating voltage V_{acc} to the actual beam potential V_b [CB93]. V_b is usually called the *beam energy*, corresponding to the electron energy at the entrance of the resonator.

$$\Delta V = |V_{\mathrm{acc}} - V_b| \approx -60\,\mathrm{V} \cdot \frac{I_b/\mathrm{A}}{\beta_\parallel} \ln\left(\frac{R_R}{R_b}\right) \frac{\ln\left(R_b/R_i\right)}{\ln\left(R_R/R_i\right)}. \qquad (1.17)$$

The radii of the coaxial insert and the cavity are given as R_i and R_R, the beam current as I_b. The average radius of the beam at the cavity center is given by $R_{b,R}$, and its radial extension is omitted in good approximation. The maximum attainable beam current can be derived from the limiting value $\beta_\parallel \to 0$ as:

$$I_{\text{lim}} \approx 4\pi \, \frac{m_e \, c_0^2}{e \, Z_0} \, \frac{\gamma \sqrt{\left(1 - \sqrt[3]{1 - \beta_\parallel^2}\right)^3}}{2 \ln\left(\dfrac{R_R}{R_b}\right) \dfrac{\ln(R_b/R_i)}{\ln(R_R/R_i)}} \tag{1.18a}$$

$$\approx 17070 \, \text{A} \cdot \frac{\gamma \left(\dfrac{\beta_\parallel}{\sqrt{3}}\right)^3}{2 \ln\left(\dfrac{R_R}{R_b}\right) \dfrac{\ln(R_b/R_i)}{\ln(R_R/R_i)}} \tag{1.18b}$$

This maximum possible beam current is also called the *limiting current* I_{lim}. The voltage depression ΔV and the limiting current I_{lim} are plotted in Fig. 1.5(a) and Fig. 1.5(b) for a generic gyrotron ($R_R = 25\,\text{mm}$, $R_b = 10\text{mm}$, $V_{\text{acc}} = 80\,\text{kV}$, $I_b = 60\,\text{A}$). In adiabatic approximation ($\beta_\perp^2/B = \text{const.}$ as described in Sec. 3.1.2), the voltage drop efficiently leads to a reduction of the normalized axial velocity $\beta_\parallel$. For a conventional gyrotron with a hollow waveguide cavity, the voltage depression is much higher in comparison to a gyrotron with a coaxial cavity, as indicated at $R_i \to 0$ in Fig. 1.5(a). The limiting current is therefore much smaller for a conventional gyrotron than for a coaxial one. Because of the greater margin to the limiting current and the lower voltage depression, the coaxial gyrotron exhibits considerable advantages over conventional designs.

Magnetron injection guns can be classified into two types: In diode-type guns (Fig. 1.6(a)), the accelerating voltage V_{acc} is applied to the cathode whereas the anode and the coaxial insert are grounded. In triode-type guns (Fig. 1.6(b)), an additional modulation anode is installed opposite to the emitter. This additional degree of freedom introduced by the voltage V_{mod} applied to the modulation anode provides the possibility to sensitively adjust the electric field $\vec{E}_E$ in the vicinity of the emitter. The variation of V_{mod} allows a better control over the beam properties, especially the velocity ratio α.

The region in the gyrotron tube between the electron gun and the interaction cavity is called the *beam tunnel*. Any RF-oscillation inside this section

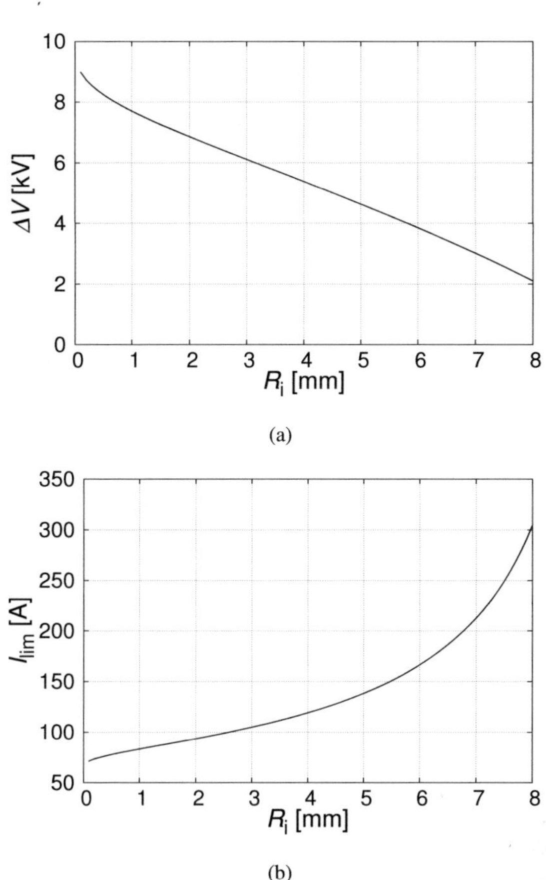

Fig. 1.5.: Dependence of the voltage depression (a) and the limiting current (b) on the radius of the coaxial insert R_i (for beam parameters see text)

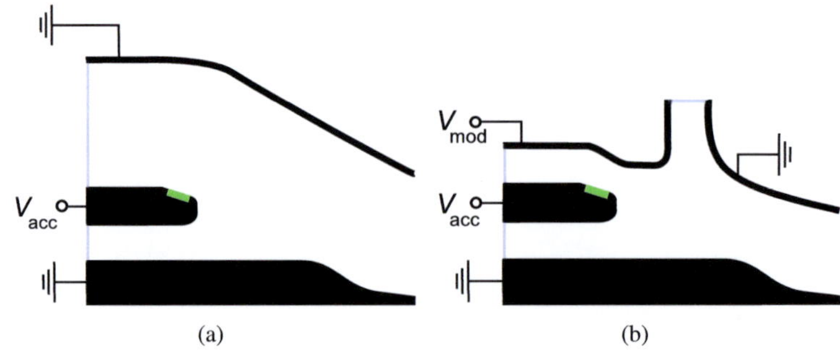

(a)   (b)

Fig. 1.6.: Simplified schematics of (a) diode- and (b) triode-type magnetron injection guns

has to be avoided in order to allow an undisturbed electron beam to reach the desired interaction region. In the beam tunnel the magnetic field is strongly inhomogeneous and the diameter of the beam is reduced to the desired diameter for interaction. To avoid parasitic oscillations, absorbing materials and/or adequate geometrical setups are used to damp possible resonant structures. The risk of unwanted resonances increases significantly with higher beam energies and the beam tunnel design becomes challenging in high-power gyrotrons [GDF+10].

1.3.3. Coaxial cavity

In the cavity the energy exchange between the helical electron beam and the RF-field takes place due to the electron-cyclotron-interaction. The cavity in coaxial high-power gyrotrons is usually a three-section structure as shown in Fig. 1.7(a): An input downtaper and a uniform middle section followed by an output uptaper. The transitions between each section are smoothed using parabolic functions.

The longitudinal structure of the field inside the cavity can be described by a complex profile $f(z)$. Fig. 1.7(b) displays the longitudinal field $f(z)$ of an operating mode along the cavity with the maximum at the cylindri-

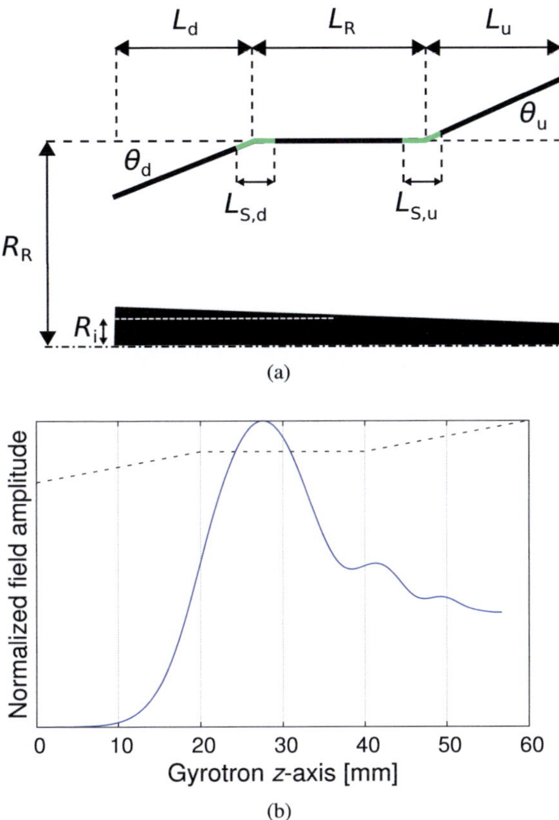

Fig. 1.7.: Axis-symmetric sketch of a coaxial cavity with a tapered coaxial insert (a) and typical longitudinal field profile (b) inside the cavity

cal cavity section. The transverse structure of the field is given by the eigenmode of the waveguide oscillation in the resonator [Bor91]. Fig. 1.8 shows examples for possible TE-eigenmodes scaled to the corresponding frequency.

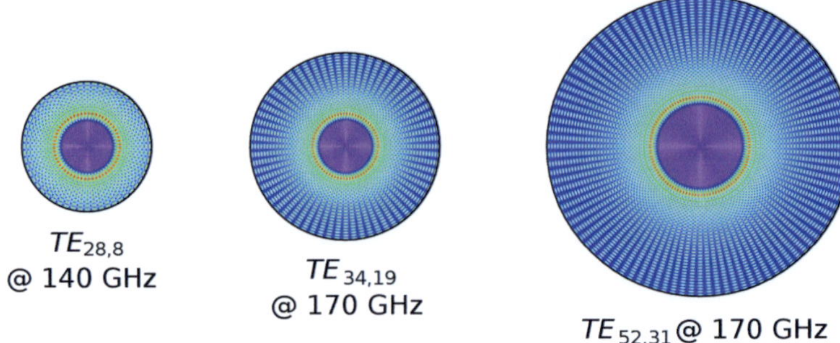

Fig. 1.8.: Distribution of the transverse electric field of several eigenmodes (nonrotating)

For a cylindrical waveguide cavity, the eigenmodes are given as $TE_{\mathrm{m,p,n}}$, $TM_{\mathrm{m,p,n}}$-modes and, especially for a coaxial structure, the TEM-mode. For $TM_{\mathrm{m,p,n}}$ modes the transverse field components E_φ and E_r vanish in a long cavity with a resonance close to cutoff. As a result, no efficient energy exchange with the transverse velocity component of the electrons is possible. The TEM mode has no cutoff frequency and therefore no total reflection towards the downtaper section, which makes a resonance with high quality factor impossible. Stable and efficient gyrotron operation is only possible with high-order $TE_{\mathrm{m,p,n}}$ modes, because their transverse field components do not vanish, and the dimensions of the resonant cavity are not limited to the scale of the emitted radiation's wavelength. The third axial index n describes the number of field maxima in the z-direction and is usually chosen equal to 1 in high-power gyrotrons. For better readability the index n is neglected in the following notation.

The *quality factor* or Q-factor is a dimensionless parameter which is defined as the ratio of $2 \cdot \pi$ times the total stored energy W_{tot} within an resonator divided by the energy, which is released per period. Simplified, the quality factor describes how much damping appears in an oscillator or resonator. The total quality factor Q_{tot} of a gyrotron cavity can be split into an ohmic Q_{ohm} and a diffractive portion Q_{dif}. Accordingly, the power P_{elec}, which is converted from the transverse electron velocity component, can be split into a portion P_{dif}, which is radiated out of the cavity, and a portion for the loading which describes the dissipated ohmic losses P_{ohm}. One can show that:

$$\frac{1}{Q_{\text{tot}}} = \frac{1}{Q_{\text{dif}}} + \frac{1}{Q_{\text{ohm}}} \tag{1.19}$$

For high-power gyrotrons and the main operating mode, Q_{dif} is usually dominant, because most of the released energy is radiated out of the cavity. An approximate analytical expression is:

$$Q_{\text{dif}} = \frac{4\pi K}{n^2} \left(\frac{L_{\text{R}}}{\lambda_0}\right)^2 \tag{1.20}$$

In this case, Eq. (1.20) indicates that Q_{dif} is nearly independent of the chosen indices m and p, and for modes with equal longitudinal index n, Q_{dif} depends only on the wavelengths λ_0. For a homogeneous cylindrical middle section, the factor K is a constant determined by the complex reflection coefficients $\Gamma_{1,2}$ at both ends of the cavity [FN88]. The advantages and use of a tapered cavity and insert are described later in this section.

$$\frac{K}{n} \approx \frac{1}{1 - |\Gamma_1| \cdot |\Gamma_2|} \tag{1.21}$$

Q_{ohm} can be estimated using the skin depth $\delta = \sqrt{2/(\mu_0 \omega \sigma)}$ and the effective conductivity σ of the material [VZO76].

$$Q_{\text{ohm}} = \frac{R_{\text{R}}}{\delta} \left[1 - \left(1 - \frac{m^2}{\chi_{\text{m,p}}^2}\right)\right] \tag{1.22}$$

R_{R} is the cavity radius as given in Fig. 1.7(a) and $\chi_{\text{m,p}}$ is the eigenvalue of the mode. For $TE_{\text{m,p}}$ modes $\chi_{\text{m,p}}$ is the p-th root of the first derivative of the m-th Bessel function ($J_{\text{m}}'(\chi_{\text{m,p}}) = 0$). The eigenvalues for TM-modes are given as the roots of the Bessel function. The m-th Bessel function and its roots can be calculated using [AS72]:

$$J_{\mathrm{m}}(\chi_{\mathrm{m,p}}) = \frac{(-j)^m}{2\pi} \int_{-\pi}^{+\pi} \cos(m\varphi)e^{j\chi_{\mathrm{m,p}}\cos(\varphi)}d\varphi = 0 \qquad (1.23)$$

In the following, all considerations are usually given for TE-modes unless otherwise noted.

The ohmic wall loading ρ_{R} of the cavity with a length L_{R} of the cylindrical middle section and an area of $S = 2\pi R_{\mathrm{R}} L_{\mathrm{R}}$ is given as:

$$\overline{\rho}_{\mathrm{R}} = \frac{P_{\mathrm{ohm}}}{S} = \frac{Q_{\mathrm{dif}} P_{\mathrm{dif}}}{Q_{\mathrm{ohm}} S} \approx \frac{2\sqrt{\pi}}{c_0^3 \sqrt{\mu_0 \sigma}} \cdot \frac{f^{2,5} Q_{\mathrm{dif}}}{\frac{L_{\mathrm{R}}}{\lambda_0}\left(\chi_{\mathrm{m,p}}^2 - m^2\right)} P_{\mathrm{dif}} \qquad (1.24)$$

The resonance frequency is assumed to be close to cutoff ($\chi_{\mathrm{m,p}}/R_{\mathrm{R}} \approx 2\pi/\lambda_0$), as it is usually the case in high-power gyrotrons.

Eq. (1.24) describes a decreasing wall loading of the cavity for modes with higher $(\chi_{\mathrm{m,p}}^2 - m^2)$. The possibility to operate the gyrotron in a high-order mode allows the size of the cavity to be much larger than the emitted radiation's wavelength. The challenge which has to be dealt with is the dense mode spectrum at higher $\chi_{\mathrm{m,p}}$-values. It is desirable to achieve stable mono-frequency oscillation of the main operating mode without disturbance due to competing modes. The difficulty is to reach an efficient operating point, which is robust within an acceptable parameter range.

An approximate inner boundary for the electromagnetic field in the cavity cross-cut is called the *caustic radius R_{c}*, which is slightly smaller in comparison to the radius of the azimuthal field maximum as indicated in Fig. 1.8.

$$R_{\mathrm{c}} = R_{\mathrm{R}} \frac{|m|}{\chi_{\mathrm{m,p}}} \qquad (1.25)$$

At radii smaller than R_{c} the field amplitude decays rapidly. For first harmonic cyclotron interaction, the maximal coupling between electron beam and field is located close to the azimuthal field maximum. To reduce mode competition, high-power gyrotrons are operated using a radially small beam, which is optimally aligned for the desired operating mode. The optimal beam position is determined, among other considerations, by utilizing a maximal coupling of the electron beam to the RF-field.

As shown in Sec. 1.3.4, it is possible to define different rotations of the mode relative to the helical motion of the electrons. In the modes' notation this is reflected by different signs of the azimuthal index $\pm|m|$. If the rotation of the mode is in the same direction (negative phase term, $TE_{-|m|,p}$) as the rotation of the electrons, the maximal coupling to the electron beam is shifted to slightly smaller radii. In addition, the maximal coupling of $TE_{-|m|,p}$ modes to the electron beam is slightly higher which makes them preferable. The main competing modes for a hollow waveguide cavity are $TE_{|m|-3,p+1}$ and $TE_{|m|-2,p+1}$. Further important competing modes are the so-called azimuthal neighbors $TE_{m\pm1,p}$. For high eigenvalues, their resonance frequency separation Δf becomes smaller.

$$\frac{\Delta f}{f_{m,p}} \approx \frac{\pi}{2\chi_{m,p}} \tag{1.26}$$

Since a smaller frequency separation increases mode competition problems, it gets more difficult to obtain efficient single mode operation in a gyrotron at higher eigenvalues.

In hollow waveguides, $\chi_{m,p}$ for one mode is constant. According to Eq. (1.20), if the geometry and the longitudinal index n are fixed, Q_{dif} in a hollow waveguide resonator is mainly dependent on the frequency and not on the mode (assuming the reflections at the cavity are the same for the different modes). In general, Q_{dif} increases approximately with f^2.

This changes for coaxial cavities: If the caustic radius R_c of the mode in a coaxial cavity is close to the radius of the insert R_i, $\chi_{m,p}$ depends on the ratio C of the cavity radius R_R to the radius of the coaxial insert R_i.

$$C = \frac{R_R}{R_i} \tag{1.27}$$

For a coaxial cavity the presence of the insert has different influence on modes with different caustic radii, which can be utilized for the suppression of unwanted competitors. To get the desired change in the quality factor it is necessary to taper the coaxial insert towards the exit of the cavity (as indicated in Fig. 1.7(a)) and to introduce a so-called impedance corrugation, which consists of longitudinal grooves. Fig. 1.9 shows a coaxial insert with corrugation and its relating geometrical parameters. The width of the slots of the corrugation is small in comparison to the wavelength λ_0.

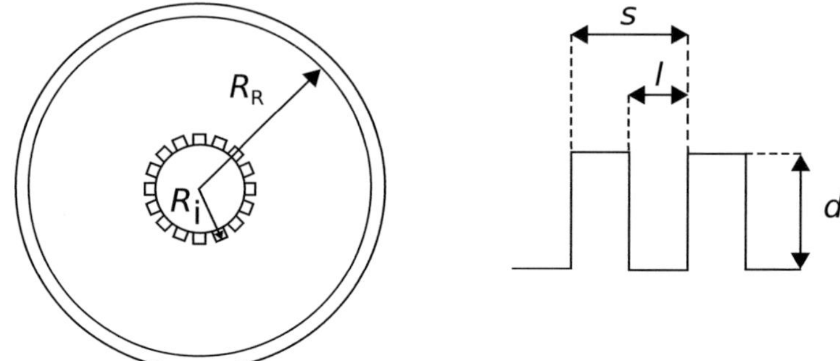

Fig. 1.9.: Sketch of the coaxial insert with longitudinal impedance corrugation

In the following the main facts for the corrugation on the coaxial insert are summarized, a detailed description can be found in [Ker96]. Fig. 1.10 shows a typical dependence of $\chi_{m,p}$ on C for modes under the influence of the coaxial insert with (blue line) and without (black line) impedance corrugation. For $C \to \infty$ the $\chi_{m,p}(C)$ curve reaches the limiting value for a hollow waveguide.

Without corrugation, a region (between the red lines in Fig. 1.10) occurs in which the $\chi_{m,p}(C)$ curve has a positive slope. For modes within this area, the eigenvalue $\chi_{m,p}(C)$ rises towards the cavity exit (C rises, because R_R is fixed and R_i decreases), so their cutoff-frequency $f_\perp$ increases and in consequence the quality factor Q increases as well. With the introduction of the impedance corrugation, all modes under the influence of the coaxial insert show a monotonic behavior of their $\chi_{m,p}(C)$ curves and the quality factors decrease towards the exit of the cavity. This effect can be utilized for efficient suppression of competing modes. To achieve a monotonic rise of C towards the exit of the cavity (even at its cylindrical middle section) it is necessary to apply a constant tapering to the coaxial insert's radius. In high-power gyrotrons, the radius of the inner rod has to be chosen small enough that the operating mode does not cause unacceptable wall losses on the inner rod, which results in a rod radius below the operating mode's caustic radius. Thus the operating mode is usually not affected by the described effect. But within this constraint, the insert's radius has to be chosen as large as possible, in a way that as many competing modes as possible experience a reduction of their quality factor.

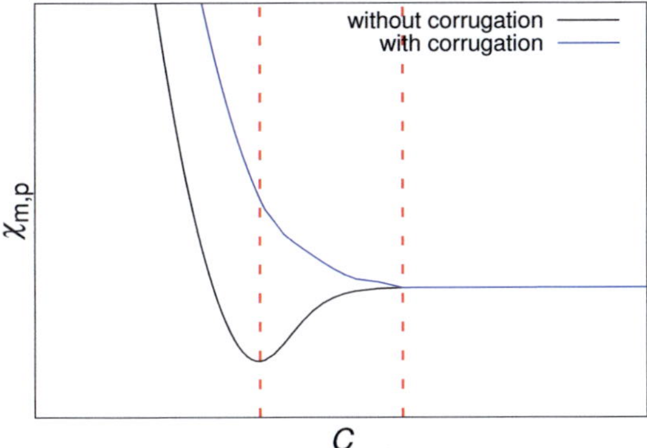

Fig. 1.10.: Typical $\chi_{\mathrm{m,p}}(C)$ curve of a TE-mode under the influence of the coaxial insert

The design process of a coaxial cavity can be divided into several steps. For the initial optimization process of the cavity its geometry is simplified using linear up- and downtapers. Afterwards the output radius of the cavity has to be fitted to the input of the quasi-optical output launcher. This is usually achieved using a nonlinear uptaper, which guarantees low mode conversion. The linear downtaper is also replaced by a nonlinear one to guarantee low mode conversion and thus low transmission of RF-power to the electron gun region and to reduce mode conversion by the whole cavity geometry.

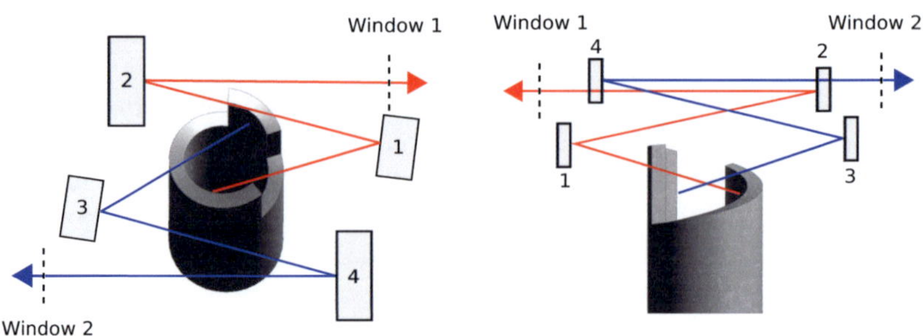

Fig. 1.11.: Possible layout for a two-beam quasi-optical output system with two mirrors per beam

1.3.4. Quasi-optical output system

After the electron-cyclotron-interaction in the cavity, the generated RF-field and the electron beam propagate within a hollow waveguide towards the so-called *quasi-optical output system*. In the quasi-optical output system, several methods, which can be described equivalent to quasi-optical phenomena, are utilized to transform the generated high-order cavity mode into one or more beams.

The quasi-optical output system has two main functions: The spent electron beam and the generated RF-power have to be separated. In addition, the high order cavity mode has to be converted to one or more linearly polarized paraxial beams, which can exit the tube radially through one or more output windows.

The quasi-optical output system of a gyrotron with two output beams is displayed in Fig. 1.11. It can be divided into two main parts: An antenna, which converts the generated high-order cavity mode into one or more radiated beams, and several reflectors (mirrors) which further change and optimize the desired beam properties.

The antenna is in principle a hollow waveguide with one or two helical cuts at its end. Usually for gyrotrons, this first part of the output system is called *launcher*. Starting from the high-order waveguide mode, a defined mode mixture is generated along the launcher using specific inner wall perturbations [BD04]. This mode mixture consists of neighboring modes with different orders and is individually optimized for the operating mode and its corresponding frequency. Multi-frequency operation of a gyrotron requires surface perturbations of the launcher which are suitable for the conversion of different operating modes [PAG$^+$09]. The mode mixture results in a focused RF-field towards the launcher exit and at each cut one directed beam is radiated into free space. To enable the desired propagation of the beams along the mirrors, the eigenmode of an arrangement of reflectors is utilized. This eigenmode of a mirror line is usually the fundamental Gaussian mode $TEM_{0,0}$. Thus, the directional radiation characteristics of the launcher are optimized to deliver highly focused beams with a Gaussian field profile.

After radiation into free space, further conversion and optimization of the beams is achieved by utilization of a mirror system. Optimized curvature radii and surface profiles for every reflector allow selective modifications on the beam properties. Each Gaussian beam can be further transformed with a different beam waist or corresponding alignment. Depending on the curvature of each mirror, the beams can be more or less focused. For example, the orthogonal beam axes can be transformed separately in order to convert an elliptical beam into a beam with circular cross section. The size of each reflector is determined by the necessary field distribution on its surface. The desired functions and properties of the reflectors and the complete mirror system can be considered analog to lenses in geometrical optics.

The profile of the first reflector for every beam is usually quasi-elliptical to transform the radiated field into a paraxial wave beam. The following mirrors are utilized to shift the waist of each Gaussian beam close to the window plane where the beam exits the tube. Due to the applied surface shape, the reflectors within the mirror system can be divided into different kinds: Mirrors which are only used to focus the beam have a surface shape which can be described by quadratical functions. So-called phase-correcting mirrors have a more complex surface shape in order to achieve specific improvements for the desired beam properties [Mic98].

In the case of high-power gyrotrons, every output beam leaves the tube through a synthetic diamond window, which acts as a vacuum barrier. The synthetic diamond is characterized by low RF-losses and very favorable thermo-mechanical properties. The diamond discs are usually produced in a chemical vapor deposition (CVD) process. In a typical CVD process, the diamond grows on a substrate surface from a hydrocarbon gas mixture (H_2 and CH_4) [TAH$^+$01]. Today, CVD diamond windows for gyrotrons can handle up to 2 MW of transmitted RF-power. Consequently for a 4 MW gyrotron, the high-order cavity mode has to be transformed into two Gaussian beams which exit the gyrotron through two output windows.

1.3.5. Collector

After electron-cyclotron-interaction in the cavity and separation from the generated microwave in the quasi-optical output system, the spent electrons are absorbed in the collector. Usually gyrotrons have an interaction efficiency of approximately 35%, which means that a large fraction of the overall kinetic energy remains in the velocity components of the electrons. The addition of an energy recovery system can significantly increase the overall efficiency up to 50% and above. A so-called *depressed collector* consists of one (single-stage) or more (multi-stage) electrodes which are biased to different voltages to recover the residual electron energy. The applied depression voltage V_{depr} can be chosen as high as possible, as long as no reflected electrons occur.

Usually pure copper or copper alloys which sustain high thermal and mechanical stresses are chosen as the collector material. The absorbed electrons apply a high wall loading on the inner surface of the collector, which is cooled from the outside. Inter alia, efficient cooling can be achieved using *hyper-vapotron* techniques. The cooling water flows perpendicular to grooves along the collector's outer wall. Inside the grooves, the water vaporizes with high turbulence, resulting in a high heat exchange coefficient at its phase change. The thickness of the collector wall has to be a trade-off: On the one hand, a thin wall leads to a high heat exchange coefficient. On the other hand, a thicker wall is necessary for sufficient mechanical stability due to the collector's proper weight and vacuum tightness with respect to recrystallization processes within the material. In addition, electrons which hit the surface knock out secondary electrons and their influence on the wall loading has to be considered. At high energy levels, several generations of secondary electrons are emitted, which might reach parts of the collector structure that are insufficiently cooled, or electrons may travel back to the resonator region where they may have a negative influence on the gyrotron interaction.

The wall loading of the collector can be decreased significantly using external magnetic field coils, which sweep the beam in longitudinal and transverse direction along the collector. Details about the several sweeping techniques can be found in Sec. 5.1. The applied sweeping frequency f_{sw} is limited by eddy currents in the collector wall. The skin depth δ of the magnetic field generated by the collector sweeping system has to be greater than the thickness d_{col} of the collector wall.

$$f_{sw} < \frac{1}{\pi \sigma \mu_0 d_{col}^2} \tag{1.28}$$

An additional reduction of the wall loading can be achieved by advanced alignment of the collector's wall relative to the electron beam. If the beam hits the surface at a lower angle the applied heat flux density is reduced. The challenge in the collector design is to optimize the magnetic sweeping system and the surface layout in order to reach a minimum average wall loading which is acceptable for long lifetime of the tube.

2. Mode selection and coaxial cavity design

The ohmic wall loading on the surface of the coaxial cavity represents one of the main restrictions in the development of high-power gyrotrons for heating and current drive of magnetically confined fusion plasmas. Due to the available cooling and material techniques the loading can not exceed values of $2\text{-}3\,\mathrm{kW/cm}^2$ in CW operation. The increased resonator size and therefore an operation with a suitable high-order $TE_{\mathrm{m,p}}$-mode can overcome this issue and make high-power long-pulse operation possible. As designs are scaled to high-power and high-frequency, the mode density in the cavity increases and it is not possible to find modes which are widely isolated from their nearest neighbors. For gyrotrons operating with high-order volume modes, the dense mode spectrum at high eigenvalues delivers a vast number of possible mode candidates and competitors.

In this chapter a mode selection process for a $4\,\mathrm{MW}$ $170\,\mathrm{GHz}$ coaxial-cavity gyrotron is shown and stable operating parameters are elaborated. To facilitate the design of the necessary two-beam output launcher and the suitability for multi-frequency operation, the chosen mode has to fulfill several additional criteria, which are considered in the procedure. A suitable geometry for the cavity including nonlinear tapers is optimized to guarantee reliable and efficient electron-cyclotron-interaction.

2.1. Theory of self-consistent multi-mode simulation

The simulation of the electron-cyclotron-interaction in the cavity is performed by deriving analytic expression based on Maxwell's equations and the Lorentz force, and by numerically solving these equations in one dimension. Reviews about the early history of gyrotrons can be found in [FGPY77] and [HG77], theoretical approaches and summaries are gathered among many others in [Edg99] and [Bor91]. Following the expansion of theory to coaxial geometries as shown in [Ker96], this section gives a rough overview of the corresponding formulas and the intention is not to be exhaustive.

Starting from Maxwell's equation, with general assumptions for homogeneous and isotropic materials, the time-harmonic ($e^{j\omega t}$) $\vec{E}$-field is described by the following differential equation:

$$\Delta \vec{E} - \frac{1}{c_0^2}\frac{\partial^2}{\partial t^2}\vec{E} = \mu_0 \frac{\partial}{\partial t}\vec{j} + \nabla \frac{\rho}{\varepsilon_0} \tag{2.1}$$

The electric field $\vec{E}$ can be split into eigenwaves $\vec{e}_k = (\vec{e}_k^{\,+} + \vec{e}_k^{\,-})/2$ in forward (-) and backward (+) direction at a given local resonator cross section.

$$\vec{E} = \sum_k \left(f_k^+(z,t)e^{jk_{\parallel,k}z}\vec{e}_k^{\,+} + f_k^-(z,t)e^{-jk_{\parallel,k}z}\vec{e}_k^{\,-} \right) e^{j\omega_k t} \tag{2.2}$$

k is an iterative number and its significance is described later. The eigenvectors $\vec{e}_k$ have the following characteristics:

$$\int_A \vec{e}_{k'}\vec{e}_k^{\,*}\,da = \delta_{k'k} \quad \text{with} \quad A : \text{cross section} \tag{2.3}$$

$$\Delta_\perp \vec{e}_k + \frac{\omega_{\perp,k}^2}{c_0^2}\vec{e}_k = 0 \tag{2.4}$$

The transverse Laplace operator is given as $\Delta_\perp = \Delta|_{\partial/\partial z=0}$.

$f_k^\pm$ describes the time-depending complex field amplitudes. The eigenwaves in forward $\vec{e}_k^{\,-}$ and backward $\vec{e}_k^{\,+}$ direction only differ in the sign of their z-component. As a consequence, the field profiles $f_k^\pm$ can be split into the transverse $f_{\perp,k}$ and longitudinal $f_{\parallel,k}$ components. For better readability the transverse component $f_{\perp,k}$ is simply written as f_k.

$$f_{\mathrm{k}} = \frac{1}{2} \cdot \left(f_{\mathrm{k}}^{+}(z,t)e^{jk_{\parallel}z} + f_{\mathrm{k}}^{-}(z,t)e^{-jk_{\parallel}z} \right) \tag{2.5}$$

$$f_{\parallel,\mathrm{k}} = \frac{1}{2} \cdot \frac{k_{\perp,\mathrm{k}}^{2}}{k_{\parallel,\mathrm{k}}^{2}} \left(f_{\mathrm{k}}^{+}(z,t)e^{jk_{\parallel}z} + f_{\mathrm{k}}^{-}(z,t)e^{-jk_{\parallel}z} \right) \tag{2.6}$$

The expressions are only valid close to cutoff, which is usually the case for gyrotrons. Within the following sections, the general electron-cyclotron-interaction is differentiated from the specific interaction within gyrotrons close to cutoff. As mentioned before, efficient interaction in the cavity is only possible with TE-modes because the transverse field components are not equal zero. TE-modes do not have longitudinal electric field components and can therefore be described solely with the transverse field portions f_{k}.

The equation for the field profile in the resonator without simplifications is given in Eq. (2.7). The first line of this formula contains the terms for the field profile of the eigenvectors, the second and third line the coupling due to mode conversion and the fourth line the excitation and feedback from the electron beam.

$$\frac{\partial^{2}}{\partial z^{2}}\left(f_{\mathrm{k}}(z,t)\right) + \frac{\omega_{\mathrm{k}}^{2} - \omega_{\perp,\mathrm{k}}^{2}}{c_{0}^{2}} f_{\mathrm{k}}(z,t) - \frac{1}{c_{0}^{2}}\frac{\partial^{2}}{\partial t^{2}} f_{\mathrm{k}}(z,t) - j\frac{2\omega_{\mathrm{k}}}{c_{0}^{2}}\frac{\partial}{\partial t} f_{\mathrm{k}}(z,t)$$

$$+ \sum_{\mathrm{k}'} \left[\left(2\frac{\partial}{\partial z}\left(f_{\mathrm{k}'}(z,t)\right) \int_{A} \frac{\partial}{\partial z}(\vec{e}_{\mathrm{k}'}) \cdot \vec{e}_{\mathrm{k}}^{*}\, da \right. \right.$$

$$\left. \left. + f_{\mathrm{k}'}(z,t) \int_{A} \frac{\partial^{2}}{\partial z^{2}}(\vec{e}_{\mathrm{k}'}) \cdot \vec{e}_{\mathrm{k}}^{*}\, da \right) e^{j\omega_{\mathrm{k}'}t} \right] e^{-j\omega_{\mathrm{k}}t}$$

$$= \mu_{0} \left[\frac{\partial}{\partial t} \sum_{\mathrm{n}} \left(\frac{\vec{u}_{\mathrm{n}} \cdot \vec{e}_{\mathrm{k}}^{*}}{u_{\mathrm{z,n}}} \cdot I_{\mathrm{b,n}} \right) \right] e^{-j\omega_{\mathrm{k}}t} + \int_{A} \nabla\frac{\rho}{\varepsilon_{0}} \cdot \vec{e}_{\mathrm{k}}^{*}\, da \cdot e^{-j\omega_{\mathrm{k}}t}$$

$$\tag{2.7}$$

The current density can be obtained using a summation over n macro-particles (ensemble of electrons) with the complex relativistic momentum

$\vec{u}_\mathrm{n}$ and the longitudinal component $\vec{u}_\mathrm{z,n}$. With several assumptions the equation for the field profile f can be simplified. The three major simplifications are the negligence of mode conversion, of space charges and of fast time dependent changes ($\partial^2/\partial t^2 = 0$).

$$\frac{\partial^2}{\partial z^2}\left(f_\mathrm{k}(z,t)\right) + \frac{\omega_\mathrm{k}^2 - \omega_{\perp,\mathrm{k}}^2}{c_0^2} f_\mathrm{k}(z,t) - j\frac{2\omega_\mathrm{k}}{c_0^2}\frac{\partial}{\partial t} f_\mathrm{k}(z,t)$$
$$= \mu_0 \left[\frac{\partial}{\partial t}\sum_\mathrm{n}\left(\frac{\vec{u}_\mathrm{n}\cdot\vec{e}_\mathrm{k}^*}{u_\mathrm{z,n}}\cdot I_\mathrm{b,n}\right)\right] e^{-j\omega_\mathrm{k}t} = R_\mathrm{k} \tag{2.8}$$

The equation for stationary conditions of the field profile can be obtained with $\partial/\partial t = 0$.

$$\frac{\partial^2}{\partial z^2}\left(f_\mathrm{k}(z)\right) + k_{\parallel,\mathrm{k}}^2 f_\mathrm{k}(z) = \mu_0\left[\frac{\partial}{\partial t}\sum_\mathrm{n}\left(\frac{\vec{u}_\mathrm{n}\cdot\vec{e}_\mathrm{k}^*}{u_\mathrm{z,n}}\cdot I_\mathrm{b,n}\right)\right]\cdot e^{-j\omega_\mathrm{k}t} = R_\mathrm{k}$$
$$\tag{2.9}$$

An additional negligence of the excitation term $R_\mathrm{k} = 0$ results in the "cold" field profile of the current free cavity [VZO$^+$69]. In other words, no feedback from the electron beam is considered.

$$\frac{\partial^2}{\partial z^2}\left(f_\mathrm{k}(z)\right) + k_{\parallel,\mathrm{k}}^2 f_\mathrm{k}(z) = 0 \tag{2.10}$$

The mathematical description for the electron movement are based on the equilibrium of the relativistic Lorentz- and the centrifugal-forces for the undisturbed rotating particles.

$$\frac{\partial\gamma m_\mathrm{e}\vec{v}}{\partial t} = q(\vec{E} + \vec{v}\times\vec{B}) \tag{2.11}$$

q is the particle's charge and $q = -e$ for electrons. The velocity component is substituted by the electron momentum $\vec{u}$.

$$\vec{u} = \frac{\gamma\vec{v}}{c_0} \tag{2.12}$$

$\vec{u}$ can be split into a longitudinal $\vec{u}_z$ and a transverse $\vec{u}_\perp$ part and the angular position around the guiding center is described by the phase ψ.

$$\vec{u} = \vec{u}_\perp + \vec{u}_z = u_r\vec{e}_r + u_\varphi\vec{e}_\varphi + u_z\vec{e}_z$$
$$= u_\perp\cos(\psi - \varphi)\vec{e}_r + u_\perp\sin(\psi - \varphi)\vec{e}_\varphi + u_z\vec{e}_z \quad (2.13)$$

ψ can be divided into a fast changing portion $\Omega_f t$, with an angular velocity Ω_f close to the cyclotron frequency, a slow variable phase Λ and a phase offset ξ. To describe the bunching and energy exchange effect with the RF-field, the slow variable phase Λ and the amplitudes of the velocity components are the necessary parameters which have to be calculated.

$$\psi = -\Lambda(t) + \Omega_f t + \xi \quad (2.14)$$

The differential equation for the transverse electron movement can be written after introduction of the complex transverse momentum p.

$$p = u_\perp e^{-j\Lambda} \quad (2.15)$$

$$\frac{\partial}{\partial t}(p) = -\frac{e}{m_e}\left(\frac{E_r}{c_0} + j\frac{E_\varphi}{c_0} - \frac{u_z(B_\varphi - jB_r)}{\gamma}\right)e^{-j(\Omega_f t + \xi - \varphi)}$$
$$+ j\left(\frac{e}{m_e}\frac{B_z}{\gamma} - \Omega_f\right)p \quad (2.16)$$

The longitudinal momentum is given as:

$$\frac{du_z}{dt} = -\frac{e}{m_e}\left(\frac{E_z}{c_0} + u_\perp\frac{\cos(\psi - \varphi)B_\varphi - \sin(\psi - \varphi)B_r}{\gamma}\right) \quad (2.17)$$

Eq. (2.16) and (2.17) can be further simplified using several assumptions: The voltage depression is considered indirectly by lowering the electron energy at the cavity input. $E_z = 0$, which only introduces errors for strong mode conversion and this is usually not the case in the gyrotron cavity. All fast time-variant magnetic fields are neglected, because the interaction is only considered with TE-modes close to cutoff and the transversal magnetic field components are small in comparison to the external field. Also the relatively small static magnetic field of the electron beam is neglected. Inside the cavity it can be assumed that the external magnetic field is nearly constant and consists only of the z-component. The exact position of one particle can be determined by integration over its velocity and the derivation over time can be transformed to one over z. The simplified equations for the particle movement without tapering of the magnetic field are:

$$\frac{c_0 u_z}{\gamma} \frac{d}{dz}(p) + j\left(\Omega_f - \frac{\Omega_0}{\gamma}\right) p = -\frac{e}{m_e c_0}(E_r + jE_\varphi)\, e^{-j(\Omega_f t + \xi - \varphi)} \tag{2.18}$$

$$u_z = \text{constant.} \tag{2.19}$$

Eq. (2.18) can be solved parallel to the equations for the field (Eq. (2.8) or (2.9)). In [Ker96] computer programs for the numerical solution of these time-dependent self-consistent equations have been developed. For the studies in this work these codes and formulas are used respectively. To summarize, within self-consistent calculations, the field equations and the equations for the motion of the electrons are solved simultaneously including time-dependence and multi-mode scenarios.

2.2. Technical and physical requirements

For the design of a gyrotron with specific frequency and output power several physical and technical requirements have to be met. Tab. 2.1 lists the design constrains for a $4\,\text{MW}$ $170\,\text{GHz}$ gyrotron, which is suitable for long-pulse operation. The values represent conservative choices considering long-lifetime and reliability.

An additional design parameter is the *Fresnel coefficient* C_F, which describes the losses due to diffraction in a quasi-optical resonator with a given length L_R.

Parameter	value
Frequency f_0 [GHz]	170.0
Peak ohmic wall loading at cavity ρ_R [kW/cm^2]	< 2.0
Peak ohmic wall loading at insert ρ_i [kW/cm^2]	< 0.2
Electric field at emitter E_E [kV/cm]	< 70.0
Emitter current density j_E [A/cm^2]	< 5.0
Maximum limiting current I_b/I_E	≈ 0.5
Maximum emitter radius $R_{E,max}$ [mm]	< 100.0

Tab. 2.1.: Design constrains for a 4 MW 170 GHz CW gyrotron

$$C_F = \frac{\pi}{4} \frac{\left(\frac{L_R}{\lambda_0}\right)^2}{\sqrt{\chi_{m,p}^2 - m^2}} \tag{2.20}$$

For the gyrotron design, C_F is a measure for the deformation of the phase front in a tapered waveguide due to diffraction. In former designs $C_F \geq 1$ should verify that a single mode description is appropriate. This constrain is used with reduced significance due to the low analogy to the description of quasi-optical resonators. To satisfy the phenomenological importance of the Fresnel coefficient its value is chosen as high as possible.

The main constraints for CW gyrotrons are the peak ohmic wall loading of the cavity ρ_R and the coaxial insert ρ_i. A cavity generating P_{dif} of power in one mode with a surface material conductivity σ and surface size $S = 2\pi R_R L_R$ has an ohmic wall loading ρ_R of:

$$\rho_R = \frac{P_{ohm}}{S} = \frac{Q_{dif}P_{dif}}{Q_{ohm}S} = \frac{1}{2\sigma\delta}|H_{tan}|^2 \tag{2.21}$$

Using Eq. (2.21), the components of the corresponding normalized eigenvectors from Eq. (2.2) and the magnetic field values, the wall loading of a homogeneous waveguide can be calculated. The magnetic field components $H_{r,k}$, $H_{\varphi,k}$ and $H_{z,k}$ are obtained using the second Maxwell equation

Eq. (4.1) and Eq. (2.2). Assuming time harmonic and symmetrical conditions, the wall loading on the surface of a coaxial cavity is given as:

$$\rho_{\mathrm{R}} = \frac{\delta C_{\mathrm{k}}^2}{4\omega\mu_0} \left(J_{\mathrm{m}}(\chi_{\mathrm{k}}) - \frac{J_{\mathrm{m}}'(\chi_{\mathrm{k}})}{N_{\mathrm{m}}'(\chi_{\mathrm{k}})} N_{\mathrm{m}}(\chi_{\mathrm{k}}) \right)^2$$
$$\cdot \left(\frac{m^2}{R_{\mathrm{R}}^2} \left| \frac{\partial}{\partial z} f_{\mathrm{k}}(z,t) \right|^2 + \frac{\chi_{\mathrm{k}}^4}{R_{\mathrm{R}}^4} |f_{\mathrm{k}}(z,t)|^2 \right) \tag{2.22}$$

C_{k} is a constant for normalization and given in Eq. (6.8).

The loading on the different walls of the grooves on the coaxial insert are described by several models. One approach is the *surface impedance model* (SIM) (summarized formulas in [Ker96]), the other one is based on a singular integral equation (SIE) [DZ04]. The main differences are that SIE predicts significantly lower losses for a given field inside the cavity and the loading on the top surface of the corrugation is not equal zero for a depth $d = \lambda/4$ of the grooves in comparison to SIM. Details for both models are described in Sec. 6.2.1. In self-consistent calculations as utilized within this work, the surface impedance model is implemented to calculate the wall loading on the different walls of the corrugation. Within SIM, the loading on the side walls of the grooves decreases from a fixed value ρ_{bottom} on the bottom side of the grooves, with a $\cos^2$-characteristic towards their top side. The amplitude of the normalized field components inside the grooves with a depth d is proportional to $1/\cos(k_{\perp,\mathrm{k}}d)$. For a specific depth $d = \lambda_{\perp}/4$ this expression has to vanish, otherwise the amplitude of the field inside the grooves would become infinite. The ohmic wall loading ρ_{top} on the top side of the grooves can therefore be reduced significantly, in contrast to the loading ρ_{bottom} on the bottom areas. If the radius of the coaxial insert gets close to the caustic radius of the operating mode, its ohmic wall loading increases rapidly. A sufficiently high safety margin is necessary between the largest required radius of the coaxial insert for suppression of competing modes and the increase of the wall loading. In summary, with an impedance corrugation, the loading on the outer side of the insert can be decreased, but the loading on the bottom side of the grooves gains importance.

2.3. Formalism of normalized variables

One possibility to describe the nonlinear theory of a gyrotron oscillator and to consider the quality and efficiency of the beam-wave interaction for a large number of TE-modes is the use of universal pendulum equations [DT85]. These generalized differential equations describe the evolution of the electron energy and their phase with respect to the RF-field. Accordingly, the so-called *method of normalized variables* [KDST85] allows the evaluation of the efficiency of the electron-cyclotron-interaction for a broad mode spectrum. Summaries of the theory can be found in [Iat96]. [Bor91] gives an overview on different systems of dimensionless systems and variables. The method had verified success in several gyrotron experiments and is simple to adapt to various problems.

With several assumptions and simplifications the terms for the relative transverse energy u of a single electron can be given as:

$$\frac{du}{d\zeta} = 2Ff\zeta\sqrt{1-u}\sin(\theta) \tag{2.23}$$

$$\frac{d\theta}{d\zeta} = \Delta - u - 2\frac{Ff(\zeta)}{\sqrt{1-u}}\cos(\theta) \tag{2.24}$$

In Eq. (2.23) and (2.24) the three free normalized variables are: The field amplitude F in the cavity, the detuning parameter Δ and the interaction length μ.

$$F = \frac{E_0 \beta_\perp^{s-4}}{B}\left(\frac{s^{s-1}}{s! \cdot 2^{s-1}}\right) J_{m\pm s}(k_\perp R_b) \quad \text{with} \quad k_\perp = \frac{\chi_{m,p}}{R} \tag{2.25}$$

s denotes again the cyclotron harmonic and the algebraic sign of the Bessel function J_m corresponds to the two possible rotations of the RF-field.

$$\mu = \pi \frac{\beta_{\perp,0}^2}{\beta_{\parallel,0}} \frac{L}{\lambda} \tag{2.26}$$

$$\Delta = \frac{2}{\beta_{\perp,0}^2}\left(1 - \frac{s\omega_c}{\omega}\right) \tag{2.27}$$

In addition, $L = 2/k_\parallel$ is the effective cavity length and ζ the normalized axial position.

$$u = \frac{2}{\beta_{\perp,0}^2}\left(1 - \frac{\gamma}{\gamma_0}\right) \quad , \quad \zeta = \pi\frac{\beta_{\perp,0}^2}{\beta_{\parallel,0}}\frac{z}{\lambda} \tag{2.28}$$

In the case of an axial Gaussian field profile $f(\zeta)$

$$f(\zeta) = e^{-(2\zeta/\mu)^2} \tag{2.29}$$

the perpendicular interaction efficiency is:

$$\eta = \frac{\gamma_0 - \gamma}{\gamma_0 - 1} = \frac{\gamma_0\beta_{\perp,0}}{2(\gamma_0 - 1)}\eta_\perp \quad \text{with} \quad \eta_\perp = \langle u(\zeta_\text{out})\rangle_{\theta_\text{in}} \tag{2.30}$$

If F and μ are fixed, the third parameter Δ can be chosen to give highest possible efficiency. $\eta_\perp$ is the average value over an adequate number of electron input phase positions θ_in. Fig. 2.1 shows the areas of constant efficiency $\eta_\perp$ after optimization of Δ [Ker96]. For an assumed Gaussian field profile the transverse efficiency $\eta_\perp$ has its maximum of 72% at $\mu = 19$ and $F = 0.13$.

2.4. Mode pre-selection process

The maximal possible output power for a coaxial cavity can be estimated using:

$$P_\text{dif} \leq \frac{\rho_{R,\max}\left[1 - \left(\frac{m}{\chi}\right)^2\right]}{f^{2.5}4\pi\left(\frac{L_R}{\lambda_0}\right)\frac{Q_\text{diff}}{Q_\text{min}}}\frac{c_0^3\sqrt{\mu_0\sigma}}{2.8\sqrt{\pi}}\chi^2 \tag{2.31}$$

The notation χ (without indices) denotes a non-discrete value, for which in a dense mode spectrum adequate close equivalents $\chi_{m,p}$ can be found. Eq. (2.31) shows, if the ratio m/χ is fixed, the possible output power for coaxial resonators is dependent on χ^2. In comparison, in hollow wave-guide cavities the output power is approximately linearly dependent on χ.

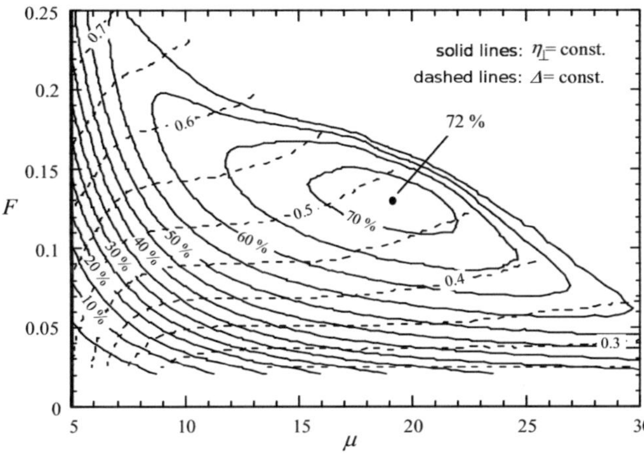

Fig. 2.1.: Possible efficiency $\eta_\perp$ in F-μ-plane after optimization of Δ

The higher possible output power within coaxial geometries is even more evident considering the advanced mode selection due to a possible corrugation of the cavity's outer wall.

For given beam parameters the output power can be determined and the compliance with technical constrains can be checked consequently. Within [Ker96] the program *MAXPO* was developed based on the theory of normalized variables to calculate the maximum possible gyrotron output power considering the physical requirements and efficiencies. Fig. 2.2 shows the calculated highest output power over the continuous Bessel root χ for different normalized resonator lengths μ with respect to the constraints from Tab. 2.1. Within this estimations no mode conversion is considered.

The highest output power is predicted for χ values between 150 and 170. Within this range a high number of possible mode candidates is available. An additional constrain for the chosen mode is a suitable spread angle ψ to facilitate the design of the necessary two beam launcher. For a two-beam launcher with a five-fold perturbation on the launcher wall (see Ch. 4) the spread angle of the chosen mode should be close to $\psi = 360°/5 = 72°$.

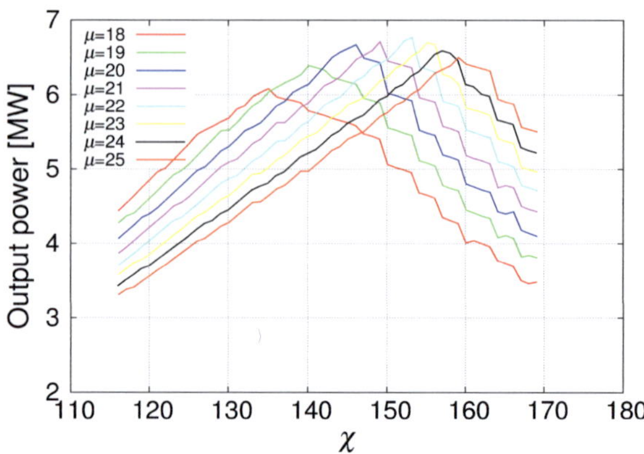

Fig. 2.2.: Achievable output power versus χ for different normalized interaction lengths μ

For the modes close to the desired spread angle, cavity geometries are optimized and possible operating points are identified. For every mode candidate a design is carried out by computing the interaction efficiency in self-consistent stationary single-mode approximations for various parameters until an acceptable cavity design compatible with the design goals, such as efficiency, Q, wall losses, output power, etc., is obtained.

With a specific safety margin (10%) concerning internal losses and stray radiation the selected mode should deliver around 4.4 MW in the cavity. Within $150 \leq \chi \leq 170$, 34 modes can be identified in single mode and stationary operation to deliver ≈ 4 MW at 170 GHz with a spread angle of $72° \pm 1.0°$. Within a range of $72° \pm 0.5°$, 12 modes are present. The choice of accuracy for the spread angle is arbitrary, but with respect to the high number of possible modes adequate. Tab. 2.2 summarizes the collected data for the most promising mode candidates.

Mode	$\chi_{m,p}$	ψ [°]	P_{out} [MW]	η [%]	ρ_R [kW/cm^2]	ρ_i [kW/cm^2]
$TE_{-46,28}$	150.79	72.24	4.45	32.14	2.13	0.13
$TE_{-47,28}$	152.11	72.00	4.90	34.40	2.17	0.12
$TE_{-48,28}$	153.43	71.77	4.48	33.56	2.08	0.15
$TE_{-49,28}$	154.75	71.54	4.33	35.05	2.02	0.14
$TE_{-48,29}$	156.74	72.17	4.48	35.18	1.95	0.12
$TE_{-49,29}$	158.06	71.94	4.54	35.00	1.81	0.13
$TE_{-50,29}$	159.38	71.72	4.52	32.01	1.88	0.15
$TE_{-50,30}$	162.68	72.10	4.27	32.87	1.75	0.14
$TE_{-51,30}$	164.00	71.88	4.62	34.90	1.95	0.19
$TE_{-50,31}$	165.98	72.47	4.47	32.13	1.89	0.18
$TE_{-52,31}$	168.63	72.04	4.42	33.44	1.94	0.15
$TE_{-53,31}$	169.95	71.83	4.43	32.21	1.93	0.16

Tab. 2.2.: Mode candidates with $150 \leq \chi \leq 170$ and $71.5° \leq \psi \leq 72.5°$

2.5. Self-consistent calculation and cavity optimization

In a next step the identified mode candidates from Sec. 2.4 are evaluated in transient self-consistent multi-mode simulations. The program *SELFT* [Ker96] is utilized and the corresponding theoretical basics are given in Sec. 2.1. The goal is to find one mode which is suitable to deliver the desired output power within a broad range of operating parameters. For every candidate parameter studies are performed and possible operation points are investigated. Within this procedure the mode $TE_{-52,31}$ showed very promising behavior concerning reliable output power, mode competition, efficiency and stability.

The parameters for an optimized cavity for the $TE_{-52,31}$ mode with linear input and output tapers and the length of the parabolic smoothings can be found in Tab. 2.3. With the chosen radius of the cavity R_R equal to 47.32 mm an output frequency of 170.15 GHz, slightly higher than the desired 170.00 GHz are obtained. Due to the expected thermal expansion of the cavity, as determined in Sec. 6.1, the output frequency of the heated cavity will be very close to the desired one. The applied wall loading on the coaxial insert increases rapidly when its surface approaches the operating's mode caustic radius. Therefore the radius of the coaxial insert is chosen with a safety margin in respect to the wall loading to guarantee sufficient cooling but still high reduction of the competing modes' Q-values.

Parameter	Value
$L_d/L_R/L_u$ [mm]	30.0/16.5/22.0
$\theta_d/\theta_R/\theta_u$ [°]	2.5/0.0/2.2
Smoothing $L_{s,d}/L_{s,u}$ [mm]	$2\times2.2/2\times2.0$
$R_R/R_i/R_b$ [mm]	47.32/13.00/15.10
θ_i [°]	-1.0

Tab. 2.3.: Optimized cavity geometry for the $TE_{-52,31}$ mode (for parameters see Fig. 1.7(a))

The dense mode spectrum with the coupling to the electron beam normalized to the mode $TE_{-52,31}$ is displayed in Fig. 2.3(a) (for $R_b = 15.1\,\mathrm{mm}$). Fig. 2.3(b) shows the components of the normalized field profile of the possible operating mode $TE_{-52,31}$ using self-consistent stationary simulations. At lower z-values, at the downtaper, the wave is in cutoff and the maximum field amplitude is located in the middle of the cylindrical cavity. After optimization the parameters for the operating point, the wall loading and the Q-value of the resonator are given in Tab. 2.4. The parameters for the wall loading are assumed with a factor nearly equal two, for real dispersion strengthened copper in comparison to the ideal material. Detailed considerations about the wall loading are given in Sec. 6.1.1 or [Gei91].

Parameter	Value
P_{out} at cavity [MW]	4.42
U_b [kV]	132
I_b [A]	105
B_R [T]	7.36
α	1.32
R_b [mm]	15.10
η incl. ohmic losses [%]	33.4
ρ_R [kW/cm^2]	1.94
ρ_i [kW/cm^2]	0.15
Q_{diff}	2870

Tab. 2.4.: Cavity parameters and operating point after optimization for the $TE_{-52,31}$ mode

Up to now the cavity's outer wall is assumed smooth without corrugation. The possibility to achieve stable and efficient interaction using the mode $TE_{-52,31}$ in an even broader parameter range is discussed in Sec. 2.8.

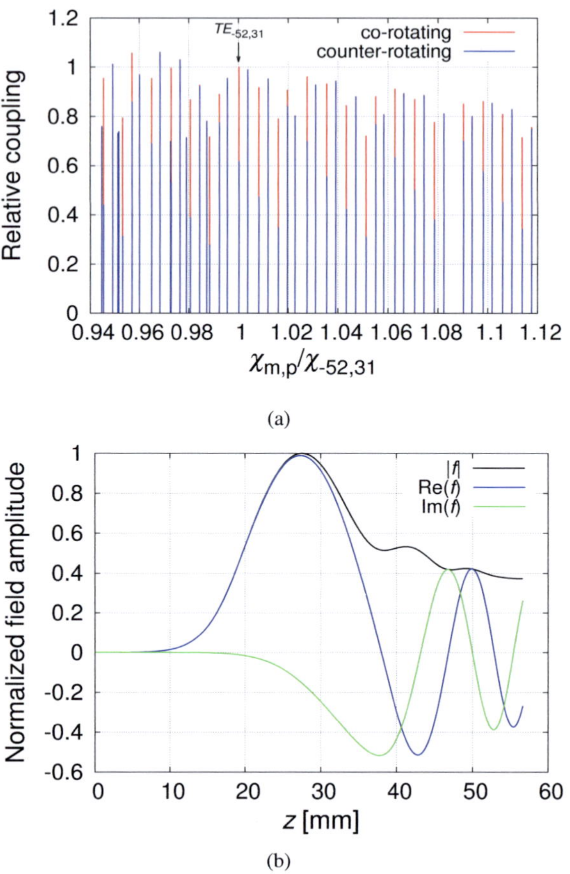

(a)

(b)

Fig. 2.3.: Normalized spectrum (a) and normalized self-consistent stationary field components (b) for the $TE_{-52,31}$ mode with $R_\mathrm{b} = 15.1\,\mathrm{mm}$

2.6. Introduction of nonlinear tapers

The optimization of the coaxial cavity in Sec. 2.5 is obtained using linear input and output tapers and parabolic smoothings in between. In a next design step, nonlinear tapers are introduced to fit the cavity geometry to the beam tunnel and launcher layout. These nonlinear tapers have to guarantee low power transmission towards the gun and beam tunnel region, lowest possible mode conversion towards the launcher and throughout the whole geometry. Fig. 2.4 shows the simplified cavity geometry with three positions for possible mode conversion. In addition, the layout of the tapers has to be as short as possible due to the opening beam of the spend electrons along the magnetic field lines towards the launcher.

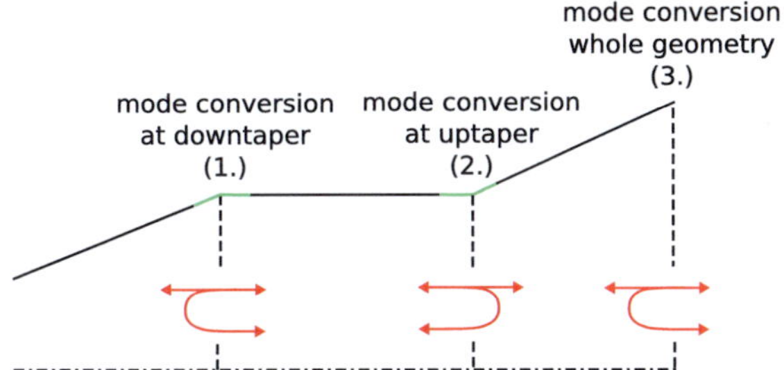

Fig. 2.4.: Simplified cavity geometry with three possible locations for mode conversion

The synthesis of the nonlinear tapers is performed using a scattering matrix program based on eigenwave expansion [Hoe94]. For TE-modes with an azimuthal index m the orthonormal eigenfunctions for coaxial waveguides with radius R can be written as:

$$T(r,\varphi) = \tau \left(N'_\mathrm{m}(k_\mathrm{r} R_\mathrm{i}) J_\mathrm{m}(k_\mathrm{r} r) - N_\mathrm{m}(k_\mathrm{r} r) J'_\mathrm{m}(k_\mathrm{r} R_\mathrm{i}) \right) \sin(m\varphi) \quad (2.32)$$

with the scaling factor τ

$$\tau = \left\{ \varepsilon_0 \frac{\pi}{2} \left[\left((k_r R)^2 - m^2 \right) Z_m^2(k_r R) \right. \right.$$

$$\left. \left. - - \left((k_r R_i)^2 - m^2 \right) \left(\frac{2}{\pi k_r R_i} \right)^2 \right] \right\}^{\frac{1}{2}} \tag{2.33}$$

and

$$Z_m(k_r R) = N'_m(k_r R_i) J_m(k_r R) - N_m(k_r R) J'_m(k_r R_i) \tag{2.34}$$

Eq. (2.32) is valid for $E_z = 0$ (TE-modes) and an ideal conducting waveguide, where the field parallel to the surface vanishes $\vec{E}_\parallel = 0$. With $T(r, \varphi)$ every arbitrary tangential field distribution for TE-modes, which is adequate to the boundary conditions, can be derived.

$$F = \sum_{k=1}^{\infty} \sum_{l=1}^{\infty} \xi_{k,l} T_{k,l} \tag{2.35}$$

The coefficient ξ is the evolution factor. The decomposition is analog to a Fourier transformation, in which a given function is split into sine- and cosine-waves. For the characterization of concentric radii changes in coaxial waveguides the layout is discretized into slices with constant geometry. At the resulting joints, the field transformations are described by scattering matrices $[S]$. The length of each constant part results in a phase change of the field components. For one transition between two steps with a change of the radii the scattering matrices are given as [Pie77][Pie81]:

$$\left[S^{21} \right] = 2 \left([E][E]^T + [I] \right)^{-1} [E] \qquad \left[S^{11} \right] = [I] - [E]^T \left[S^{21} \right] \tag{2.36}$$

$$\left[S^{12} \right] = \left[S^{21} \right]^T \qquad \left[S^{22} \right] = [E] \left[S^{12} \right] - [I] \tag{2.37}$$

with

$$[E] \cdot \left(\vec{a}^1 + \vec{b}^1 \right) = \left(\vec{a}^2 + \vec{b}^2 \right) \tag{2.38}$$

$[I]$ describes the identity matrix with $[I]_{i,j} = \delta_{i,j}$. According to the concentric radii changes, coupling is only attended between modes with the same azimuthal index m.

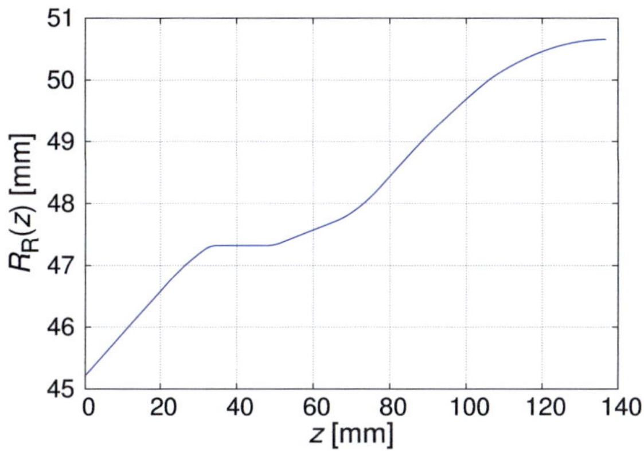

Fig. 2.5.: Complete cavity geometry with optimized nonlinear tapers

The ratio between the launcher radius at its input $R_{L,0}$ (as shown in Ch. 4) and the cavity radius R_R is called the *oversize factor*. The uptaper design is optimized for an oversize factor equal 1.07 [TYA$^+$05] resulting in $R_{L,0} = 50.633\,\mathrm{mm}$. Recent studies on possible after cavity interaction scenarios suggest higher values for this factor. The complexity of the nonlinear taper design is not dependent on the choice of the oversize factor, but the length of the taper towards the launcher itself. This might lead to serious problems regarding the opening electron trajectories. One additional magnetic coil close to the launcher, as described during the electron gun design in Ch. 3, can compensate the widening of the beam and permit a longer launcher layout.

Fig. 2.5 shows the optimized cavity profile including nonlinear up- and downtapers. The corresponding reflection coefficients can be found in Tab. 2.5, in which only values $\geq 0.01\%$ are given. The optimized geometry shows low ($< 0.1\%$) power transmission towards the gun region and low overall mode conversion ($\approx 0.3\%$). The nonlinear uptaper starting from the end of the cylindrical cavity has a length of 87.0 mm. The total length of the cavity including all tapers is 135.0 mm, which is comparable with the designs for the existing FZK/KIT 140 GHz 1 MW ($\approx$ 100 mm) gyrotron for the Wendelstein 7-X stellarator and the European 170 GHz 2 MW ($\approx$ 140 mm) gyrotron for ITER.

	Position 1	Position 2	Position 3
Reflection $TE_{-52,29}$	–	–	0.09
Reflection $TE_{-52,30}$	0.04	4.42	0.08
Reflection $TE_{-52,31}$	99.91	–	99.68
Transmission $TE_{-52,30}$	–	0.15	–
Transmission $TE_{-52,31}$	–	95.29	–
Transmission $TE_{-52,32}$	0.01	0.05	0.07

Tab. 2.5.: Mode conversion [%] at the different positions (see Fig. 2.4) for the optimized nonlinear tapers

2.7. Time-dependent multi-mode calculations

In a next step, time dependent start-up simulations are performed considering a linear rise of the accelerating voltage. The time-dependent gradient of the pitch factor α is initially chosen to be linear. After the design process of the triode-type gun in Ch. 3, the velocity ratio α is set as shown in Fig. 2.6(b). With a modulating voltage of $V_{\mathrm{mod}} = -64.4$ kV the pitch factor has a low level over a broad region and rises rapidly near the final operating point. As a result, the transverse electron velocity component is low and the risk of unwanted oscillation is reduced before the nominal operating parameters of the main mode are reached.

For reliable multi-mode calculation, the choice of competing modes, which are considered in the simulation, is crucial. The two main selection crite-

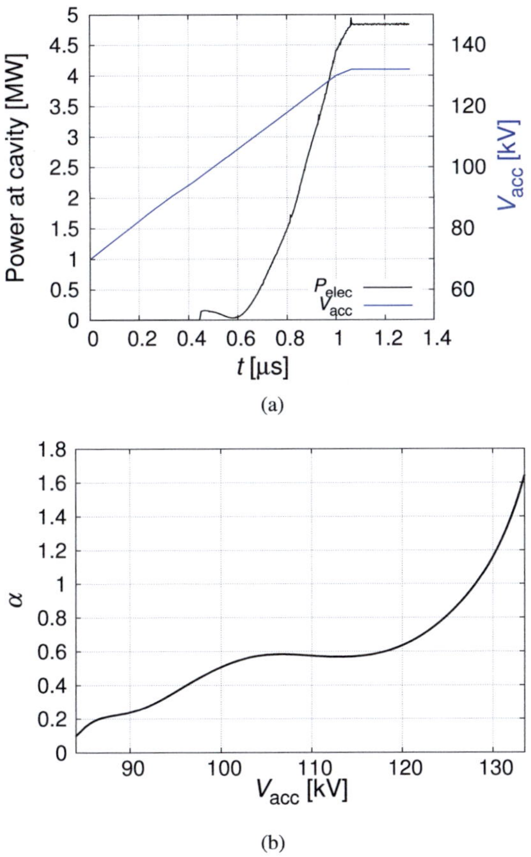

Fig. 2.6.: Output power at cavity for the $TE_{-52,31}$ mode (single-mode) and corresponding pitch factor α (for a triode-type gun with $V_{\mathrm{mod}} = -64.4\,\mathrm{kV}$)

ria for the consideration are the modes' frequency separation and the coupling factor to the electron beam normalized to the main operating mode. According to the change of the cyclotron frequency during startup of the gyrotron, due to the voltage rise to around $132\,\mathrm{kV}$ and the consequently following change in the relativistic factor γ, a frequency range $-10\,\mathrm{GHz}$ and $+20\,\mathrm{GHz}$ around the desired frequency f_0 has to be considered. The passed relativistic cyclotron frequency $f_\mathrm{c}(V_\mathrm{acc})$ for different magnetic field values is plotted in Fig. 2.7. Corresponding simulations neglecting modes with lower coupling to the electron beam related to the working mode show a necessity for the consideration of modes with a normalized coupling higher than 70%. For the given geometry, 63 modes, including the main mode, have a coupling above 70% in the frequency range $-10\,\mathrm{GHz}$ and $+20\,\mathrm{GHz}$.

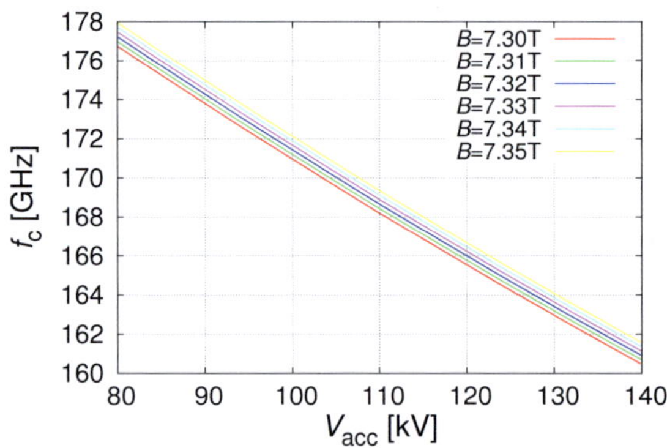

Fig. 2.7.: Relativistic cyclotron frequency f_c versus V_acc at several magnetic field values

As described above, within the self-consistent approach the coupled differential equations for the field profile and the particle movement have to be solved. The numerical implementations in [Ker96] utilize the Crank-Nicholson-Scheme (field) [PTVF92] in combination with a Predictor-Corrector-Method (particle movement) [EMR93]. The typical transit time for electrons through the cavity is around 1 ns. For several 1000 geometrical discretization steps along z and several 10 particles the equations for the electron movement need to be solved with a time step < 1 ps. The necessary time step for the field profiles is smaller than 0.2 ns. If the field profile is assumed to be slowly changing over time, the equations for the particles only need to be solved once within the time step for the profile. Descriptively, it is assumed that the excitation term R_k for the field profile is nearly constant during the transit time of the particles. It is possible to identify the greatest possible time step for fast and reliable simulation around 0.01 ns. A second important numerical parameter for reliable simulation is the number of utilized macro-particles (ensembles of electrons). To characterize the energy exchange between electron beam and field corresponding to Eq. (2.16), the average value over a sufficient number of electrons has to be obtained. For the shown simulations in this work, the number of particles considered in the simulation is chosen to be > 30. Probably the most crucial numerical parameter is the number of discretization steps for the slow variable phase Λ as described in Eq. (2.14). A safe choice for the number of discretization steps is the next higher prime number related to the highest azimuthal index m of the modes considered in multi-mode calculation. This selection reduces the possibility of beating effects and non-physical behaviors.

Fig. 2.8(a) shows the achievable output power of 4.35 MW with the selected operating mode $TE_{-52,31}$. The corresponding values for V_acc, I_b and η considering the losses in the cavity can be found in Fig. 2.8(b). η drops in comparison to single mode simulations to 33%, because the upper region of the stability area (V_acc over B plane) can not be reached due to mode competition.

In addition, Fig. 2.8(a) shows a spurious mode $TE_{+47,30}$ with a generated power of several kW parallel to the main operating mode. This mode can be identified as a backward wave in the downtaper section of the cavity. The mode $TE_{+47,30}$ has a relatively high coupling to the electron beam and an output frequency of approximately 163.4 GHz. The field profile of this specific competing mode $TE_{+47,30}$ and the main mode $TE_{-52,31}$ is shown in Fig. 2.9. With a strongly closed downtaper section it is possible to suppress the parasitic backward wave, but, as a consequence, the

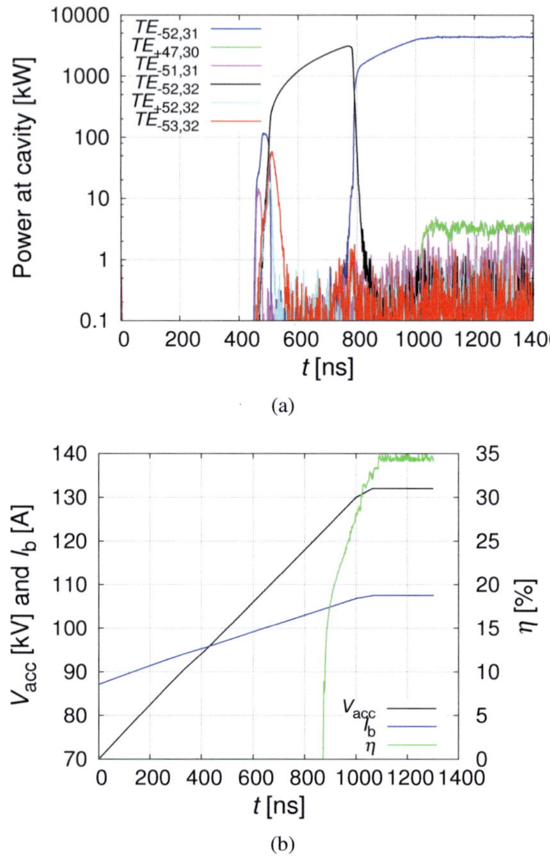

(a)

(b)

Fig. 2.8.: Output power at cavity (a), and V_{acc}, I_{b} and η (b) for the $TE_{-52,31}$ mode in multi-mode (corresponding parameters in Tab. 2.4)

desired operating mode loses efficiency and the mode conversion properties towards the gun region become insufficient. Another approach for the suppression is the introduction of a widely opened downtaper section in combination with sufficient reflection of the main mode and an optimized shape of the coaxial insert. This leads to a lower Q value for possible generated backward waves. Within the optimization process, it is not possible to find geometrical downtaper setups with the proposed characteristics. An artificial negligence of the backward wave in multi-mode simulations leads to other spurious modes with high relative coupling to the electron beam in the downtaper section. The occurrence of competing backward waves in the downtaper section can be interpreted as first evidences for the upper achievable power limit for coaxial-cavity gyrotrons. The design of the downtaper at these high beam energies is therefore a trade-off between highest possible efficiency of the main operating mode and sufficient suppression of corresponding spurious oscillations.

In summary, stable single-mode operation is possible for a sufficient range of operating parameters even when the density of modes in the cavity becomes very high. The $TE_{-52,31}$ mode is well capable to deliver the desired output power at $170\,\text{GHz}$ and is stably oscillating over a wide range of nominal parameters without disturbance from competing modes. Spurious backward oscillations slightly lower the achievable interaction capability of the main mode, but still a satisfying value for the efficiency of 33% is attainable.

2.8. Stability areas and cavity outer wall corrugation

With an additional longitudinal corrugation of the cavity outer wall competing modes can be suppressed which leads to more stable operating points and broader stability regions [Ker96] [CD84] [PFK$^+$01]. The width of the mode converting grooves is, in comparison to the impedance corrugation in the insert, not small in comparison to $\lambda/2$.

The main idea of the cavity outer wall corrugation is the coupling of the

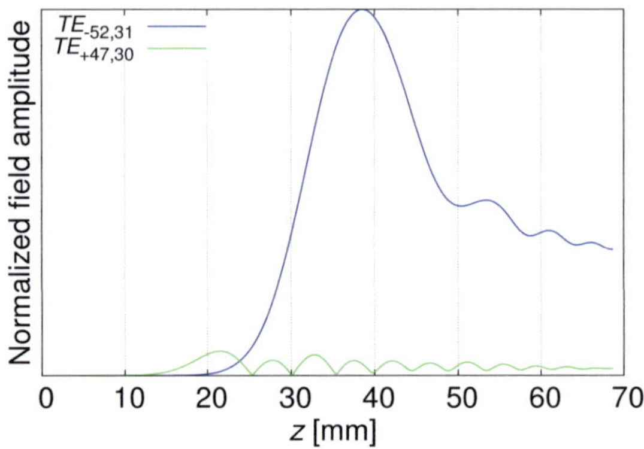

Fig. 2.9.: Field profiles for the main mode $TE_{-52,31}$ and the backward wave $TE_{+47,30}$

main competitors to corresponding degenerate modes with a relative low Q-value. The number of grooves along the cross section is Δm, and only modes with a difference in the azimuthal indices equal to Δm can be coupled ($TE_{-m,p} \rightarrow TE_{\Delta m-|m|,p}$). In addition, Δm is usually chosen to guarantee no influence of the outer wall corrugation on the main operating mode. The corrugation depth d_m dictates the strengths of coupling, which is chosen as deep as possible with respect to assure undisturbed operation of the desired mode as well. If competitors get coupled to modes with a low Q-value, an additional effect impairs their suppression: The coupling of the degenerate modes to the electron beam is usually much lower in comparison to the one of the main competing modes. Details on the mathematical theory for the cavity outer wall corrugation can be found in [Ker96].

For coaxial cavities the main competitors are usually the azimuthal neighboring modes $TE_{m\pm1,p}$ or the hollow waveguide competitors which are not adequate damped by the influence of the coaxial insert. The azimuthal neighbors have nearly the same radial structure as the main operating mode $TE_{m,p}$ and can therefore be affected only by disturbance of the azimuthal symmetry.

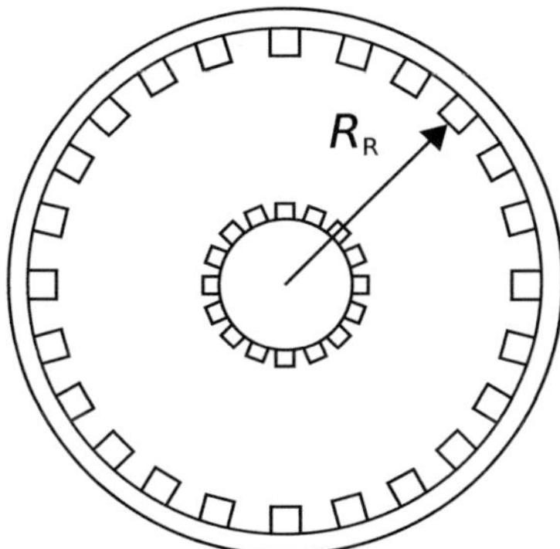

Fig. 2.10.: Axial sketch of inner and outer conductor with corrugation

Fig. 2.11(a) and 2.11(b) show the stability region for the main mode $TE_{-52,31}$ with and without azimuthal neighbors. Within simulation the competing modes $TE_{-51,31}$ and $TE_{-53,31}$ are assumed to be converted into degenerated modes corresponding to an adequately chosen Δm. The stability area with suppressed azimuthal neighbors is broadened significantly towards both, higher and lower values of the magnetic field B_R. Within the plots, a linear voltage rise is assumed and each contour represents an additional generated RF-power of 500 W. In summary, when the $TE_{-52\pm1,31}$ modes vanish, the region for stable operation of the desired mode is enhanced and stable operation over a broader parameter range is possible. With a different choice of Δm for the number of grooves within the outer cavity wall corrugation also other possible competing modes might be suppressed.

2.9. Multi-frequency operation (136 GHz / 204 GHz)

Heating and current drive in magnetically confined plasmas are usually applied at a fixed frequency due to the determined gyrotron frequency. Fre-

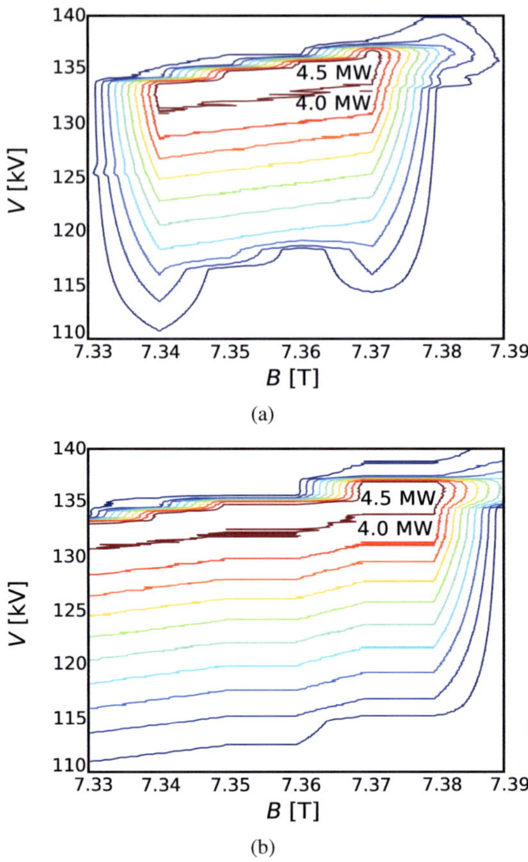

Fig. 2.11.: Stability area in the B/V_acc-plane for the $TE_{-52,31}$ mode with (a) and without (b) the azimuthal neighboring modes $TE_{-52\pm1,31}$

quency tunable gyrotrons would offer several advantages:

- The location of absorption in the plasma is adjustable by the frequency.

- The mechanical movement of the steering optics near the plasma is no longer necessary.

- A possibility of multiple operating points at different magnetic field values for tokamak and stellarator experiments.

Especially for magneto-hydrodynamic instabilities of the plasma (so-called "Neoclassical Tearing Modes"), for which the location is a priori not exactly known, frequency tunable sources would offer fast and flexible reaction (current drive) capabilities. In addition, pre-ionization of the plasma requires frequencies lower than those for current drive.

For the frequency tuning of gyrotrons several possibilities exist. Changing the magnetic field and as a consequence a mode change in the gyrotron, leads to a possible step-frequency tuning capability, which is already investigated and demonstrated [KDK$^+$00]. These types of frequency-tunable gyrotrons need an output window, which is transparent for a broad spectrum. This can be achieved using single disc windows allocated at a Brewster angle (equal $67.2°$ for synthetic diamond) relative to the RF-beam axis. A different approach using conventionally aligned output windows is a frequency change to operating points where the wavelength is equal to a multiple of $\lambda_0/2$. For example, CVD diamond windows can have a thickness of $5/2 \cdot \lambda_0$ at 170 GHz. In consequence, the transmission function for a typical disc shows the two neighboring minima at $4/2 \cdot \lambda_0$ for 136 GHz and $6/2 \cdot \lambda_0$ for 204 GHz.

The main operating mode at 170 GHz has to be chosen in a way that modes with operating frequencies at 136 GHz and 204 GHz can be found with similar caustic radii R_c to guarantee efficient electron-cyclotron-interaction and reliable capability of conversion in the quasi-optical output system [LJR$^+$10]. The perturbations of the launcher surface are optimized for one mode and a change of the caustic radius for other modes has a direct influence on the conversion efficiency.

For the chosen $TE_{-52,31}$ operating mode it is possible to find two modes with similar caustic radius at the given secondary desired frequencies. Within a future design phase this is an additional benefit of the chosen main mode to make multi-frequency operation possible. Tab. 2.6 lists the parameters of the mode candidates $TE_{-42,25}$ for 136 GHz and $TE_{-62,37}$ for 204 GHz operation. The additional modes show low deviations to the desired frequencies and caustic radii. For the higher frequency operation at 204 GHz an even higher order mode $TE_{-62,37}$ has to be utilized. Possible stable single-mode operation has to be validated within this operating regime at very higher order. In addition, the feasibility of the quasi-optical output coupler needs to be verified.

	$TE_{-52,31}$	$TE_{-42,25}$	$TE_{-62,37}$
Ideal center frequency [GHz]	170	136	204
Operation frequency [GHz]	–	136.698	203.301
Deviation of frequency [%]	–	0.51	-0.34
Ideal Bessel zero	168.629	134.903	202.355
Bessel zero $\chi_{m,p}$	168.629	135.596	201.662
Relative caustic radius $(m/\chi_{m,p})$	0.308	0.310	0.307
Deviation of caustic [%]	–	0.446	-0.299

Tab. 2.6.: Mode candidates for possible multi-frequency operation at 136 GHz and 204 GHz

3. Design of magnetron injection electron guns and the magnetic system

The gyrotron's electron gun forms the electrons emitted from the cathode into a helical beam with the properties required for effective electron-cyclotron-interaction. In the following chapter, designs for diode- and a triode-type coaxial magnetron injections guns (MIGs) including the surrounding magnetic system are elaborated. Electron guns of the magnetron injection type have proved to be very successful for the generation of the necessary annular beam in gyrotron tubes. Triode-type MIGs are equipped with a second anode, the so-called modulation anode, which introduces an additional degree of freedom to adjust the beam parameters.

At the beginning, the elementary analytical mathematical theory, limiting effects and technical constraints for the electron gun setup are described including the required flow conditions for reliable and effective interaction. An automated and parallelized optimizer with several probabilistic searching algorithms is implemented and utilized to acquire suitable diode- and triode-type MIG layouts for the $4\,\mathrm{MW}$ $170\,\mathrm{GHz}$ operation. The gun designs deliver a high quality electron beam with significant low velocity spread. Especially the triode-type design gives a novel possibility for the reduction of the radial space requirements in the warm bore-hole of the superconducting magnet. In addition, parameter studies are performed to examine the parameters' influence on several specific characteristics of the acquired beam quality. The impact of mechanical deviations from the optimum design is evaluated in order to estimate the sensitivity to permissible tolerances. The theoretical basis is mainly gathered from [Edg99] and [Ill97]. The results within [Rod09] are utilized for further optimization steps of the various designs under consideration.

3.1. Elementary mathematical theory of magnetron injection guns

The MIGs used in gyrotrons are usually operated under *temperature-limited* (TL) conditions. In contrast, in other types of microwave tubes, such as klystrons, electron guns are operated under *space-charge-limited* (SCL) emission. In TL operation, a non-zero electric field is present at the emitting surface and accelerates all emitted electrons towards the anode. In this case, the beam current is only weakly dependent on the applied voltage according to the *Schottky effect*, but it depends sensitively on the conditions of the emitting surface [Edg99]. The beam current is adjusted by the temperature of the glowing emitter and thus on the heating applied.

In the following sections, the basic theory of magnetron injection guns is given and requirements on the beam are discussed.

3.1.1. Busch's theorem

Busch's theorem puts the changes in the magnetic flux between two positions along the electrons' paths in relation to the angular electron momentum. The canonical angular momentum u_φ of a gyrating electron is described by:

$$u_\varphi = \gamma m_\mathrm{e} R^2 \frac{d\varphi}{dt} \tag{3.1}$$

R and φ are the coordinates and $d\varphi/dt$ is the azimuthal angular velocity of the gyrating electron. If the azimuthal electric field E_φ is zero and A_φ is the azimuthal component of the magnetic vector potential with $\vec{B} = \nabla \times \vec{A}$, the angular momentum u_φ, with respect to the axis of symmetry, is a constant of the motion [LL66].

$$\gamma m_\mathrm{e} R^2 \frac{d\varphi}{dt} - eRA_\varphi = \text{constant.} \tag{3.2}$$

The corresponding magnetic flux Φ_m can be described with respect to $\vec{A}$ and $\vec{B}$ within a circle of the radius R forming a surface S:

$$\Phi_{\mathrm{m}} = \int\int_{\mathrm{S}} \vec{B}d\vec{s} = \int\int_{\mathrm{S}} \nabla \times \vec{A}d\vec{s} = \oint_{\Sigma} \vec{A}d\vec{l} \qquad (3.3)$$

$d\vec{s}$ is an infinitesimal vector, whose magnitude is the area of a differential element of S, and whose direction is the surface normal. Σ is the contour surrounding the surface S, and $d\vec{l}$ is an infinitesimal vector element of the contour Σ. The magnetic flux within a circle of radius R at an axial position z is for cylindrically symmetric systems given as:

$$\Phi_{\mathrm{m}}(R, z) = 2\pi R A_{\varphi}(R, z) \qquad (3.4)$$

With the symmetry of the magnetic field inside the gyrotron cavity and the near-axis approximation $\vec{B} \approx B(z)\vec{e}_{\mathrm{z}}$, the magnetic flux Φ_{m} can be simplified for a circle with the radius R:

$$\Phi_{\mathrm{m}}(R, z) \approx R^2 \pi B(z) \qquad (3.5)$$

Consequently, for any two arbitrary positions (a) and (b) along the beam path, the ratio between the change of the angular momentum u_{φ} and the change in the magnetic flux Φ_{m} can be approximated.

$$\Delta u_{\varphi} = u_{\varphi}(b) - u_{\varphi}(a) = \frac{e}{2\pi} \left[\Phi_{\mathrm{m}}(b) - \Phi_{\mathrm{m}}(a)\right]$$
$$\approx \frac{e}{2} \left[R^2(b)B(b) - R^2(a)B(a)\right] \qquad (3.6)$$

Two positions along the gyrotron geometry can be identified where the angular momentum is zero: One is at a given radius $R = \sqrt{R_{\mathrm{b}}^2 - r_{\mathrm{L}}^2}$, because the azimuthal velocity is zero at this radius and the second one is at the starting radius R_{E} of the electron trajectories at the emitter. With these two points and Eq. (3.6), the relation between the starting radius R_{E} of the electrons and the radius of the guiding center R_{b} elsewhere on the beam path can be given using $R_{\mathrm{b}} \gg r_{\mathrm{L}}$ as it is usually the case in gyrotrons.

$$B_{\mathrm{E}}R_{\mathrm{E}}^2 = B_{\mathrm{R}}R_{\mathrm{b}}^2 \left(1 - \frac{r_{\mathrm{L}}^2}{R_{\mathrm{b}}^2}\right) \approx B_{\mathrm{R}}R_{\mathrm{b}}^2 \qquad (3.7)$$

This derivation is the basis for the definition of the magnetic compression ratio b according to Eq. (1.13).

3.1.2. Adiabatic approximation

To facilitate the description of the electron motion in crossed constant electric and magnetic fields, it is useful to separate the velocity components into the gyrating motion, with Larmor radius r_L and cyclotron frequency f_c, and a drift of the corresponding guiding centers. This approach, which is called *adiabatic approximation*, describes an infinitely slow change of the field characteristics. It is valid, if the variations of the electric and magnetic fields are rather small at dimensions comparable with the electrons' trajectories. The conditions for adiabatic motion are given as [Che95][Siv65]:

$$z_\mathrm{L}^2 \left| \frac{\partial^2 \vec{B}}{\partial z^2} \right| \ll |\vec{B}|; \quad r_\mathrm{L}^2 \left| \frac{\partial^2 \vec{B}}{\partial R^2} \right| \ll |\vec{B}| \tag{3.8a}$$

$$z_\mathrm{L}^2 \left| \frac{\partial^2 \vec{E}}{\partial z^2} \right| \ll |\vec{E}|; \quad r_\mathrm{L}^2 \left| \frac{\partial^2 \vec{E}}{\partial R^2} \right| \ll |\vec{E}| \tag{3.8b}$$

$$\frac{|\vec{E} \times \vec{B}|}{|\vec{B}|^2} \ll c \tag{3.8c}$$

During one rotation along the circle with the radius r_L, the electron travels a distance of z_L in axial direction. $|\vec{E} \times \vec{B}|$ describes the pointing vector (energy flux) and Eq. (3.8c) indicates that the energy gain over one Larmor period has to be small. Within the validity of the adiabatic approximation the magnetic moment $p_\perp^2 / 2 m_\mathrm{e} B$ is an invariant of motion.

$$\frac{p_\perp^2}{B} = \text{constant.} \tag{3.9}$$

In a region with an increasing magnetic field amplitude (and without any electric field), like the area between the gyrotron electron gun and the cavity, $p_\perp$ and $\beta_\perp$ increase and, as a consequence of energy conservation, the axial momentum $p_\parallel$ and velocity $\beta_\parallel$ decrease. Like in Eq. (3.6), it is possible to describe the relation for the transformation of the transverse velocity of the gyration motion at two arbitrary positions (a) and (b).

$$\beta_\perp(b) = \beta_\perp(a) \frac{\gamma(a)}{\gamma(b)} \left(\frac{B(b)}{B(a)} \right)^{1/2} \tag{3.10}$$

3.2. Requirements for the electron beam

In this section, the specific requirements for the gun design concerning the quality of the electron beam are discussed.

The initial set of design parameters for the electron gun can be directly derived from the specified operating mode with its output power P_{out} and desired oscillation frequency f_0. These parameters are the cyclotron frequency f_{c}, the magnetic field in the interaction cavity B_{R}, the kinetic electron energy E_{kin} (respectively the accelerating voltage V_{acc}), as well as the beam current I_{b} and the average beam radius R_{b}.

The two quantities for rating the operational stability of the MIG and to guarantee efficient electron-cyclotron-interaction are the spread of the electrons' transverse velocity from its average value $\Delta\beta_{\perp,\mathrm{rms}}$ and the spread in the beam radius $\Delta R_{\mathrm{b,rms}}$.

To describe the spread of the perpendicular velocity and the spread of the beam radius, it is advantageous to use their root mean square (rms) values. Within the root mean square, values with higher deviation have a stronger influence in comparison to the arithmetic mean. The specification for $\Delta R_{\mathrm{b,rms}}$ is analog to this definition.

A high velocity spread can significantly impair the gyrotron's performance by corrupting the mode competition and even spoil the operation [NGT04]. Many effects have influence on the magnitude of the velocity spread. Most of them are already assigned to the electrons close to the emitting surface of the gun, like thermal-spread of the initial velocities, emitter surface roughness, non-uniform space-charge effects or violation of the axis-symmetry of the electric and magnetic fields. In a second step, instabilities and space-charge effects can occur while the beam travels along its path.

According to the adiabatic approximation from Sec. 3.1.2 and neglecting additional beam effects like instabilities, the relative transverse velocity spread is in first order approximation a constant of motion along the beam path.

$$\frac{\Delta\beta_\perp}{\beta_\perp} = \text{constant.} \tag{3.11}$$

An important measure for rating the operational stability of the MIG is the maximum spread $\Delta\beta_{\perp,\mathrm{max}}$ of the electron transverse velocity component.

$$\Delta\beta_{\perp,\mathrm{max}} = \frac{\beta_{\perp,\mathrm{max}} - \beta_\perp}{\beta_\perp} \tag{3.12}$$

$\beta_\perp$ and $\beta_{\perp,\mathrm{max}}$ indicate the average and maximum values of the normalized transverse velocity. Assuming that safe operation, without reflected electrons, must be possible with electrons which have twice the maximum transverse velocity $\beta_{\perp,\mathrm{max}}$, the maximum acceptable value can be calculated from $\beta_\parallel = \sqrt{(2\beta_{\perp,\mathrm{max}})^2 - \beta^2} > 0$ to:

$$\Delta\beta_{\perp,\mathrm{max}} < \frac{1}{2}\left(\frac{\beta}{\beta_\perp} - 1\right) = \frac{1}{2}\left(\sqrt{1 + \frac{1}{\alpha^2}} - 1\right). \tag{3.13}$$

The correlation between transverse and axial velocity spread can be given for constant total energy:

$$\frac{\Delta\beta_\parallel}{\beta_\parallel} = \alpha^2 \frac{\Delta\beta_\perp}{\beta_\perp} \tag{3.14}$$

In addition, the longitudinal velocity spread $\Delta\beta_\parallel$ should be as small as possible to allow highest possible depression at the collector before reflected particles occur. Thus, the achievable overall efficiency of the gyrotron depends inter alia on the longitudinal electron velocity spread.

For efficient energy transfer from the electrons to the RF-field in the cavity, the average radius of the circular beam R_b has to be chosen to be suitable for a high coupling to the electromagnetic wave. As explained in Ch. 2, a small specific deviation of R_b can be utilized to strongly reduce the coupling of competing modes to the electron beam. The Larmor radius r_L

of the rotating electrons around their individual guiding center is given in Eq. (1.3). The radial thickness ΔR_b of the beam is not only determined by the diameter of the Larmor orbits, but also by the spatial extension of the emitter and physical effects (like space charges) which lead to a widening of the guiding centers ΔR_G, as indicated in Fig. 1.3.

$$\Delta R_\mathrm{b} = 2r_\mathrm{L} + \Delta R_\mathrm{G} \tag{3.15}$$

With reference to the definition of the velocity spread, the deviation of the average beam radius can be given as $\Delta R_\mathrm{b,rms}$. The gun design should guarantee lowest values for $\Delta R_\mathrm{b,rms}$. Another limit for the beam's thickness is given in [KDST85], as the radial extension of the guiding centers should not exceed $c_0/8f_0$ for efficient interaction with the transverse energy component.

Besides the spread of the velocity components and the beam radius, the alignment of the electron trajectories along the beam path is an additional design parameter. As illustrated in Fig. 1.4(b), the angle between the direction of the magnetic field and the emitter surface is given as φ_EB. The direction of the magnetic field can be changed by variation of the currents applied to the gun coils without modifying its magnitude and subsequently the compression ratio b and beam radius R_b. Three types of characteristic trajectories can be identified corresponding to different values for φ_EB [KLTZ92]:

- $10° < \varphi_\mathrm{EB} < 17°$ (Fig. 3.1(a)): The electrons move along intersecting trajectories. This operating condition is called *non-laminar* or *intersecting flow*.

- $24° < \varphi_\mathrm{EB}$ (Fig. 3.1(b)): In this regime, the electrons form a thick sheet without any crossing trajectories, which is called *laminar flow*.

- $17° < \varphi_\mathrm{EB} < 24°$ (Fig. 3.1(c)): The third operating domain is intermediate between laminar and non-laminar flow. All electrons pass through a narrow section (waist) without intersecting each others' trajectory. This condition is called *quasi-laminar* or *boundary-type flow*.

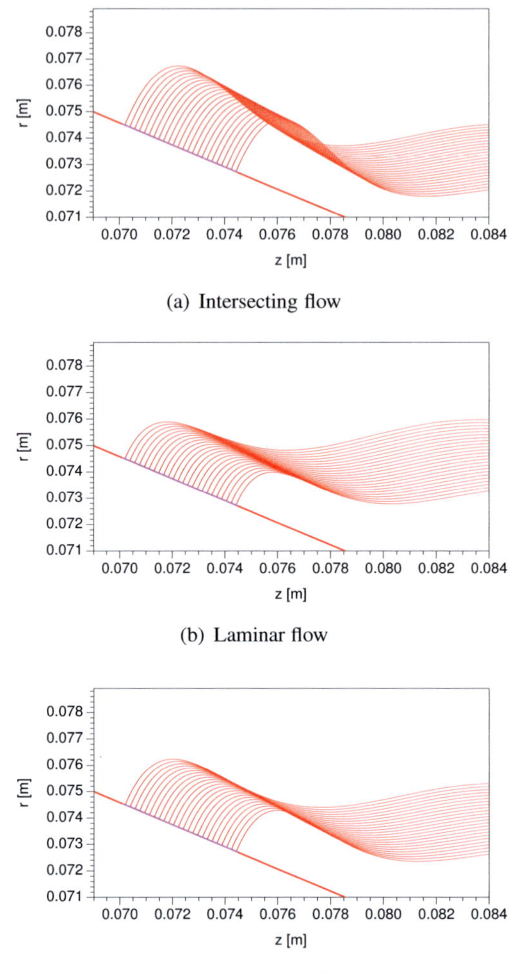

(a) Intersecting flow

(b) Laminar flow

(c) Boundary-type flow

Fig. 3.1.: Electron topologies for different angles of the magnetic field φ_{EB} starting from the emitter (pink)

In numerical simulations and practical experience, MIGs with quasi-laminar flow have delivered the best performance with respect to beam quality. As a possible explanation it is suggested that a harmonization of electrons' kinetic energies without disruption of the beam takes place at the first minimum of the radial beam width (waist of the beam) [Pio01].

An additional design requirement is the so-called *beam current neutralization*. The beam's space charge will be neutralized after an interval of several $10\,\text{ms}$ depending on the vacuum conditions in the tube. This effect slightly alters the electric field along the electron trajectories, resulting in an increase of the electron kinetic energy. The velocity ratio α will be slightly lowered because the gyrating momentum of the electrons remains constant, but their axial velocity $\beta_\parallel$ increases. With the axial velocity

$$\beta_\parallel^2 \approx 1 - \left(1 + \frac{V_\text{b}}{511\,\text{kV}}\right)^{-2} - \beta_\perp^2 \tag{3.16}$$

the drop of the velocity ratio amounts to:

$$\Delta\alpha = \beta_\perp \left[\frac{1}{\beta_{\parallel,\text{initial}}} - \frac{1}{\beta_{\parallel,\text{steady}}}\right] \tag{3.17a}$$

$$\approx \beta_\perp \left[\left(1 - \left(1 + \frac{V_\text{acc} - \Delta V}{511\,\text{kV}}\right)^{-2} - \beta_\perp^2\right)^{-1/2}\right.$$

$$\left. - \left(1 - \left(1 + \frac{V_\text{acc}}{511\,\text{kV}}\right)^{-2} - \beta_\perp^2\right)^{-1/2}\right]. \tag{3.17b}$$

For a conventional hollow waveguide gyrotron, the velocity ratio is lowered by almost 0.26 (for a nominal value $\alpha = 1.3$). In a coaxial gyrotron, this effect is significantly reduced. The velocity ratio is decreased by only approximately 0.05 for otherwise identical operating parameters. This correlation is plotted in Fig. 3.2 for an arbitrarily chosen acceleration voltage of $V_\text{acc} = 132.0\,\text{kV}$.

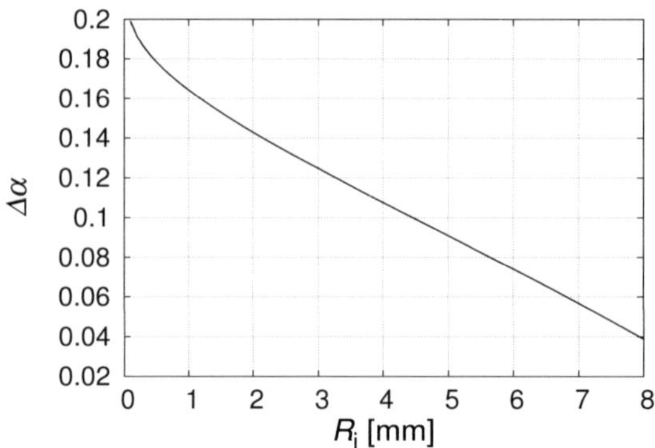

Fig. 3.2.: Dependency of the reduction of the velocity ratio α on the radius of the coaxial insert

3.3. Technical constraints and additional requirements for the gun layout

An important consideration for the design of a MIG is the maximum attainable current density j_E at the emitter surface. A too high current density can lead to severe degradations of the gun's performance, whereas for a density too low the minimum velocity spread cannot be obtained. In modern emitters, a conservative technical limit for the current density is about $5.0\,\mathrm{A/cm^2}$ to guarantee several thousands of operating hours [BHM$^+$09]. With a given beam current I_b, the current density at the surface of a ring-shaped emitter can be expressed by:

$$j_\mathrm{E} = \frac{\cos(\varphi_\mathrm{C})I_\mathrm{b}}{L_\mathrm{E}(2R_\mathrm{E} - \tan(\varphi_\mathrm{C})L_\mathrm{E})} \tag{3.18}$$

As illustrated in Fig. 3.3, the maximum radial extension of the emitter is given by R_E, its slant angle by φ_C, and the axial length by L_E. There are two ways to lower the current density: Either the length L_E is increased,

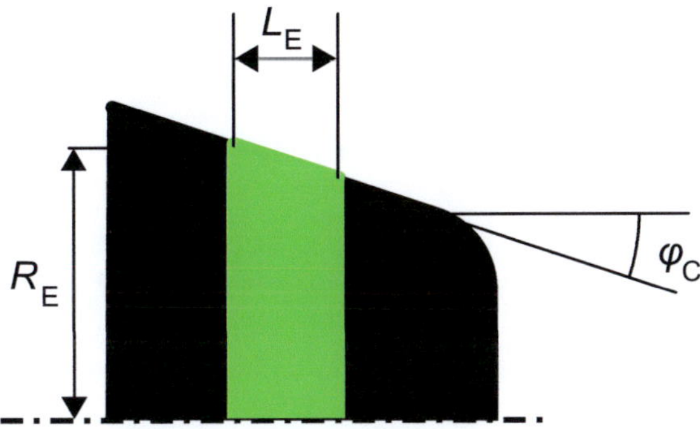

Fig. 3.3.: Axial-symmetric schematic of the cathode with the emitter region (green)

resulting in a rise of the velocity spread $\Delta\alpha_{\mathrm{rms}}$, or the radius of the emitter R_E is enlarged leading to a larger overall outside diameter of the MIG. As a consequence, a trade-off is necessary between beam quality and acceptable diameter of the warm bore-hole of the superconducting magnet required. For reliable and stable operation the maximum admissible electric field inside the electron gun has a technical upper limit.

$$E_{\max} \cdot \Delta V_{\mathrm{acc}} \leq 800\,\mathrm{kV^2/mm} \tag{3.19}$$

This expression is derived empirically and is successfully employed in various gun designs at KIT. As a result, for higher potential drops, the requirements to the electric field become more strict. This introduces an over-linear increase of the spacing required between live and grounded parts of the electron gun. A crucial design goal, as already described above, is to keep the radial dimensions of the gun as small as possible. The magnetron injection gun is mounted into the warm bore hole of the surrounding system of superconducting magnets which are operated at cryogenic temperatures. The smaller the diameter of the bore hole is, the less expensive becomes the layout of the magnet.

An additional design requirement for magnetron injection guns is the ne-

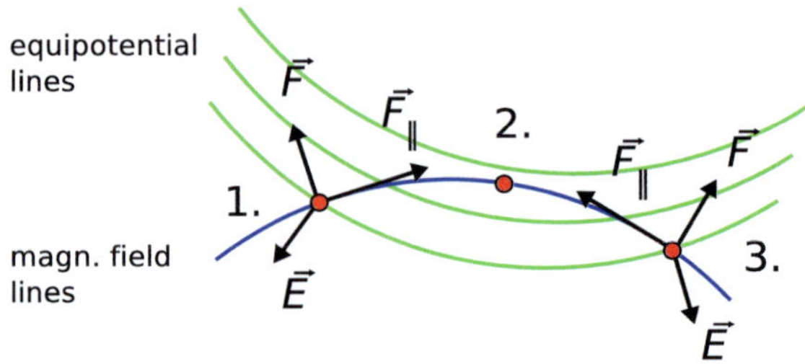

Fig. 3.4.: Electron trapping mechanism within a magnetic potential well [PHG$^+$09]

cessity for suppressing field configurations which might lead to electron trapping mechanisms. Generally there are two main trapping effects: The potential well and the magnetic mirror effect [PDD$^+$04].

Potential wells are initialized when the relative curvature of magnetic field and equipotential lines are aligned in a way that the longitudinal force $\vec{F}_\parallel$ acting on a particle changes its sign along the particle's trajectory as shown in Fig. 3.4. A negative charged particle is moving from position (1) along the magnetic field line towards the energetically favorable position (2). If the particles energy is high enough, it can leave position (2) again towards position (3). As a consequence, $\vec{F}_\parallel$ changes its direction back towards position (2) and the particle can be trapped. The accumulation of charged particles within such traps might lead to undesired effects like arcs or a limitation of the gun's high-voltage performance. This mechanism is also known as *Penning trap*. Especially in the technical (rear) part of the electron gun such disadvantageous field configurations have to be avoided.

The magnetic mirror effects occurs for some electrons which move adiabatically along the magnetic field lines towards the cavity and reach a high value of their velocity ratio α. Not only electrons which are emitted from the cathode with a high initial perpendicular velocity component can account for this effect. Particles coming from elsewhere on the cathode's surface or which are produced by a collision with molecules of the remaining gas in the tube could be trapped by this mechanism as well.

3.4. Numerical evaluation and optimization

After introduction of the mathematical basis, in the following section the geometrical setups, the utilized numerical methods and optimization routines are introduced.

3.4.1. Design specifications and simulation setup

For a 4 MW 170 GHz gyrotron one possible operating mode, namely the $TE_{-52,31}$, has been identified in Ch. 2 to guarantee effective and stable single-mode operation. The related nominal electrical parameters as well as the requirements for the electron beam are given in Tab. 3.1.

Parameter	Value
Accelerating voltage V_{acc} [kV]	132.0
Beam current I_{b} [A]	105.0
Magnetic field at the cavity B_{R} [T]	7.36
Average beam radius at the cavity R_{b} [mm]	15.10
Larmor radius at the cavity center r_{L} [μm]	106.9
Average velocity ratio α	1.3 to 1.4
R.m.s. spread $\Delta\alpha_{\mathrm{rms}}$ of the velocity ratio [%]	< 5.0

Tab. 3.1.: Nominal operating values and requirements for the electron beam

Starting from the necessary accelerating voltage V_{acc} for the $TE_{-52,31}$ mode, the maximum attainable electric field magnitude E_{max} can be calculated with Eq. (3.19) to 6.10 kV/mm. The maximum field strength of a particular design should have a sufficiently large margin to the admissible field amplitude to ensure safe and reliable operation. Based on the electric efficiency η_{elec} and the specific output power, the required initial power of the electron beam can be estimated to approximately 12-14.5 MW.

During the design of the MIGs, a limited region from the cathode to the downtaper and half of the middle part of the cavity is considered within the simulation space. The beam properties are evaluated in the middle of the cylindrical cavity section assuming the uptaper region to have negligibly low influence. As described before, the structure of the beam tunnel is intentionally complicated in order to suppress unwanted spurious oscillations. The shape of this region has, however, only a marginal influence on the beam. The complex geometry of the beam tunnel is therefore neglected within the simulation. After successful optimization of the gun design, the trajectories of the electrons have to be checked with regard to the alignment of the launcher's cuts and the mirror system.

A design version of a $165\,\mathrm{GHz}$ MIG from a former FZK experiment is utilized as a basis for the parameterization of the gun layouts [IBD$^+$97]. For the adaption to the designated operating mode $TE_{-52,31}$, the designs for the diode- and triode-layout are modified in several aspects. The corresponding geometrical parameters can be found in Fig. A.1 to A.4 in Appendix A.

The spacing between the cathode and the interaction region is increased to $410\,\mathrm{mm}$ for ensuring a smooth compression of the electron beam. The diameter of the insert at its basis is set to a high value to enhance mechanical stability and to guarantee stable operation according to the dense mode spectra at the high-order operating point. The anode's radius prior to the interaction region is narrowed to save volume. For the triode-type gun, an additional modulation anode is introduced. A slope is included to this second (modulation) anode's shape in order to provide an additional set of geometrical parameters which can be used to manipulate the electric field at the emitter during the design process. Details are explained in Sec. 3.6, and the layout of the resonator is adjusted as discussed in Sec. 2.5. The two main coils are aligned to the resonator to ensure a homogeneous magnetic induction at the interaction region. A so-called Helmholtz-setup of the main coils is not implemented, in order to generate a narrow radius of the beam only at the desired interaction region and to diminish the risk of unwanted resonances elsewhere along the beam path. The radius of the gun coils is lowered to enable sensitive control over the magnetic field at the emitter. A sufficient radial distance is necessary between gun and magnet system for the technical components of the tube and the cryostat - therefore a trade-off is required. The radius of the cathode and consequently of the emitter is kept as small as possible to guarantee a minimal overall radius of the entire electron gun. The position of the cathode is altered in a way to compress the electron trajectories approximately one Larmor period after they are extracted from the emitter. As a last step, all corners are replaced by curvatures to avoid local peaks of the electric field amplitude.

The magnetic system consists of six solenoids. Two main coils generate the required induction in the resonator and a single compensation coil nearly bucks the main magnetic field at the emitter's position. Two main coils are utilized in order to introduce an additional degree of freedom within the optimization process. These two solenoids can be merged and simplified within a later technical design phase. Two gun coils define the field at the emitter and one coil near the launcher region allows sensitive influence to the beam after the cavity. Instead of directly using the values for the currents I_{GC1} and I_{GC2} in the gun coils as parameters, derived quantities can be introduced in accordance with [Pio06]. The parameters used in the design for the operating mode $TE_{-52,31}$ are:

$$I_{\mathrm{GC1}} = \varpi \cdot \nu \cdot 107.24 \,\mathrm{A} \tag{3.20}$$

$$I_{\mathrm{GC2}} = (1 - \varpi) \cdot \nu \cdot 79.21 \,\mathrm{A} \quad \text{with} \quad 0 < \varpi < 1 \tag{3.21}$$

By changing ϖ the angle of the magnetic field φ_{B} at the emitter can be manipulated. For an arbitrary but fixed value of ϖ boundary type electron flow conditions are attained. The magnitude B_{E} can be modified by altering the factor ν respectively and $\nu = 1$ is chosen as a reference value.

The principle cathode layout is identical for the diode-type and triode-type designs. The coaxial insert is also common to all designs. The rod is assumed to be grounded because it is usually connected to the gyrotron's body potential. For the triode-type gun, the voltage of the modulation anode is given by V_{mod} and the potential of the anode is assumed to be grounded.

3.4.2. Utilized numerical implementation

Analytical expressions predicting the performance of the magnet system surrounding the tube and the MIG itself, forming the required electron beam, are not available with sufficient accuracy due to the inherent complexity. This has motivated the development of several electron-optical codes which can be used to verify designs of these systems within numerical simulations.

Several program suites, based on static self-consistent physical models, have been written in the past. Various codes have been developed specifically for gyrotron components: The EPOS-V code at the Institute for Applied Physics (IAP) in Nizhny Novgorod [Lyg95], the DAPHNE and ARIADNE++ codes at University of Lausanne (EPFL) and National Technical University of Athens (NTUA) [TWMG94], and the ESRAY code at KIT (formerly FZK) [Ill97]. All of these codes have in common that they are optimized for problems which are axis-symmetric. It is assumed that the geometry and consequently the magnetic and electric fields are not dependent on the azimuthal coordinate to save unnecessary computational efforts. The field distributions are calculated in two dimensions, whereas the electron velocities are computed in all three dimensions. The third dimension is therefore only considered partially and the codes are hence called "2.5 dimensional". In this work, the designs for the MIGs and the magnet system have been evaluated with ESRAY. This code has been successfully used to obtain MIG designs for various gyrotrons, e.g. [KAB$^+$02] and [Pio06].

The main simulation process can be subdivided into several steps:

1. The discretized Maxwell's equations are solved using a finite difference method [Yee66]. To describe the necessary curved boundaries a two-dimensional non-orthogonal mesh has to be utilized [Yee83].

2. The emitter is divided into several parts according to the grid sizing. For each section, the current density of its beamlet (the trajectory of one particle) is calculated. The total beam, which is considered in the simulation, consists out of a multitude of these beamlets.

3. Particles are extracted from the emitter assuming a uniform current density distribution as it is useful for gyrotron emitters working under temperature limited condition.

4. The field components are localized on the meshing points and need to be interpolated to the particle position.

5. The relativistic equation of motion (*Newton-Lorentz-Equation*) is solved for all particles with the locally given magnetic and electric field configuration. The particles are "pushed" through the calculation grid and a set of trajectories is obtained this way.

6. Now the space charge and current densities can be assigned to the grid points, which makes the system self-consistent.

7. To achieve the new electric and magnetic potential fields, the *Poisson equation* is solved for the acquired distribution of space charges.

8. Steps 3 to 7 are repeated until a self-consistent solution is obtained.

Details on the calculation techniques and constraints can be found in [Ill97].

3.4.3. Optimization techniques for the electron gun synthesis

Most of the geometrical parameters are strongly coupled to almost all beam properties. The modification of a single variable, which describes a shape, leads inevitably to a change of several beam properties. If a set of parameters is found for which the beam properties have a minimum deviation from the specified values, it cannot be determined whether this set is a local or global minimum. There might be another set of parameters which delivers a better beam quality than the set which was determined before. An additional aspects is that multiple design goals must be fulfilled simultaneously. This is difficult to achieve in a huge solution space (many free variables) and with strong coupling of the parameters.

To overcome this problem, optimization algorithms are utilized for the automation of the solution process. In a first step within the design of electron guns, a discrete solution space is defined with a dimension searchable within a reasonable time frame. A reduced set of parameters which describe the geometry in the vicinity of the emitter as well as the parameters which alter the electric and magnetic field at the emitter's position is selected for the solution space. For the discretization of the solution space, the minimum increment of the geometrical lengths and positions is set to the admissible mechanical tolerance of 0.1 mm. The step size of the other parameters is selected in order to generate modifications which deliver comparable effects on the beam parameters. The range of the parameters is chosen to ensure conformity to other design conditions which cannot be derived from the beam parameters obtained in the simulation. For instance, the minimum distance between cathode and coaxial insert is selected in a way that the maximum admissible electric field is not exceeded.

For optimization, the heuristic routines *simulated annealing* and *threshold acceptance* are used respectively. These algorithms can handle a huge solution space and an implemented cost function can define the significance of the several goal parameters.

The idea of the simulated annealing algorithm originated from material science [KGV83]. A hot particle is located on a rough surface. As a result of its temperature, the particle can move randomly to nearby positions which are energetically more favorable than the current position. The probability Θ that governs whether a particle executes a move or not depends on the temperature T and the energy difference ΔW between its current W_{current} and the target position W_{target} (see Fig. 3.5):

$$\Theta(W, T) = \begin{cases} e^{\left(\frac{-|W_{\text{target}} - W_{\text{current}}|}{T}\right)} & \text{if } W_{\text{target}} > W_{\text{current}} \\ 1 & \text{if } W_{\text{target}} \leq W_{\text{current}} \end{cases} \quad (3.22)$$

This function is similar to the Boltzmann-distribution. If the target position is energetically more advantageous ($W_{\text{target}} \leq W_{\text{current}}$), the move is determinately executed. If not, the exponential function in Eq. (3.22) describes the probability of the move. The more disadvantageous the energy of the new position is in comparison to the current level, the less likely the move becomes. The higher the current temperature is, however, the higher is the chance that the move occurs. A heated particle cannot therefore get stuck in local energy minima. When the particle loses temperature with progressing time, energetically advantageous positions are occupied more

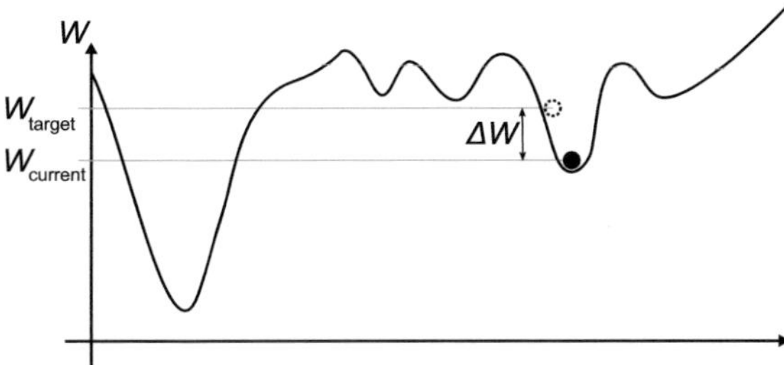

Fig. 3.5.: Example for a one-dimensional arbitrary energy function

favorably, until the temperature has reached zero and the particle will move to a local energy minimum. Whether or not the final position of the particle is close to a global minimum depends on its starting location, the initial temperature as well as the rate of cooling.

Another heuristic optimization strategy very similar to simulated annealing is threshold acceptance [DS90]. Instead of using a Boltzmann-like distribution for probabilistic determination whether a new vector is accepted, an arbitrary threshold is set which is successively lowered during simulation. If the energy difference ΔW between the current and the new solution is lower than the current threshold value, the new solution is utilized. In comparison to the simulated annealing method, less randomness is involved in the threshold acceptance algorithm. In several studies, the threshold acceptance algorithm has reached a comparably better solution than the simulated annealing method within the same number of simulation steps [LC95][SK06].

Details on the implementation and the parallelization of the optimization routines can be found in [Rod09].

3.5. Design of a diode-type electron gun

The geometry and electrical parameters of a coaxial diode-type magnetron injection gun including the necessary surrounding magnet system are opti-

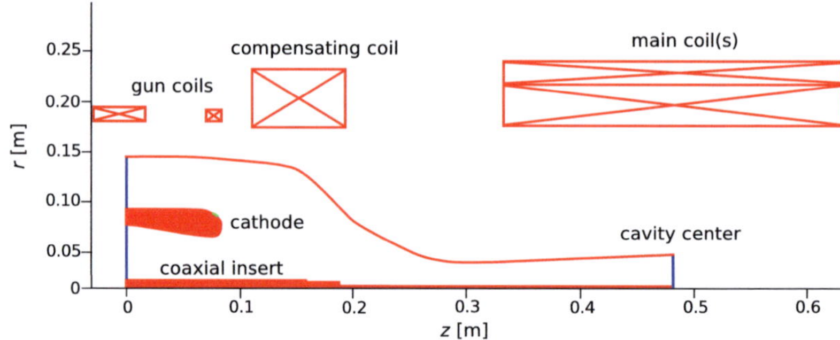

Fig. 3.6.: Axially symmetric geometry of the diode-type MIG and its magnet system

mized to deliver a high-quality electron beam for effective electron-cyclotron-interaction. The layout of the gun up to the middle of the cavity is illustrated in Fig. 3.6. The main coil is modeled as two separate windings to facilitate the optimization process with an additional degree of freedom. This simplification is no longer necessary in further technical considerations. The main parameters of the design are given in Tab. 3.2. To achieve the requested trade-off between current density and beam quality, an average emitter radius of $R_{\mathrm{E}} = 89.9\,\mathrm{mm}$ in combination with an emitter slant angle of $\varphi_{\mathrm{C}} = 21.9°$ are required. For this emitter setup an average current density of $j_{\mathrm{E}} = 4.1\,\mathrm{A/cm^2}$ is achieved for a beam current of $I_{\mathrm{b}} = 105\,\mathrm{A}$. The chosen emitter size is therefore well capable to deliver slightly higher beam currents if necessary. The characteristics of the electron beam generated by the MIG and further optimization results are listed in Tab. 3.3. The design produces a high quality electron beam and fulfills the specifications as discussed in Sec. 3.4.1. Especially the low parallel $\Delta\beta_{\parallel,\mathrm{rms}} = 1.95\%$ and perpendicular $\Delta\beta_{\perp,\mathrm{rms}} = 1.14\%$ velocity spreads for $\alpha = 1.39$ are noteworthy. The required radius R_{b} and its radial width ΔR_{b} are met and can be smoothly tuned by ϖ and ν according to Eq. (3.20) and (3.21). The velocity ratio is chosen slightly higher than the nominal value $\alpha = 1.3$ to account for the beam current neutralization as outlined in Sec. 3.2. The values for the spread of the normalized velocities are well within the requirements. The maximum spread of the transverse velocity $\Delta\beta_{\perp,\mathrm{max}}$ obtained from the simulation is far below the admissible limit. Even if additional effects (e.g. non-uniform emitter temperature) would be included into the simulation, which contribute to the velocity spread by an increase of approximately 5 % [KBT04], the maximum spread of the transverse velocity remains within acceptable limits.

For the diode-type MIG design, the velocity ratio can be adjusted by changing the angle φ_{EB} between the magnetic field vector and the emitter surface. With the parameterization given in Eq. (3.20) and (3.21), the magnetic field vector at the emitter is rotated when ϖ is modified. The reference configuration has been selected arbitrarily to give a boundary-type flow for $\varpi \approx 0.50$ (Fig. 3.8(a)). For other values, the electron trajectories become either laminar or non-laminar, resulting in a significant deterioration of the beam quality. The spread of the velocity ratio $\Delta\alpha_{\mathrm{rms}}$ is increased and the beam radius R_{b} is modified for large modifications of ϖ. The corresponding plots are given in Fig. 3.7(a) and 3.7(b). In comparison to the triode-type electron gun design, the diode layout has therefore limited adjustment capabilities.

The design is optimized in order to ensure a reasonable margin to the maximum allowable electric field ($V_{\mathrm{acc}} \approx 132\,\mathrm{kV} \rightarrow E_{\mathrm{max}} \approx 6.1\,\mathrm{kV/mm}$). By modifying the spacing between cathode and coaxial insert respectively anode, the electric fields at the insert and at the cathode are adjusted to values lower than $5.3\,\mathrm{kV/mm}$. The strength of the electric field within the design is shown in Fig. 3.8(b). Two areas with a high electric field value can be identified: One at the cathode's nose and the other on the coaxial insert. To achieve the technical design goal of the maximum allowable electric field E_{max}, a relatively huge spacing between cathode and anode is necessary. Within the diode-type design, the full accelerating voltage V_{acc} has to be applied between cathode and anode and the layout suffers large radial dimensions. As a consequence, a superconducting magnet system with a large warm bore-hole would be necessary. This problem can be overcome with a triode-type gun design as it is shown in Sec. 3.6. In summary, it is possible to find a suitable design for a diode-type magnetron injection gun which delivers a high quality beam well capable of the 4 MW 170 GHz operation point. Due to the high acceleration voltage and the corresponding high electric field values, the design is of large radial size.

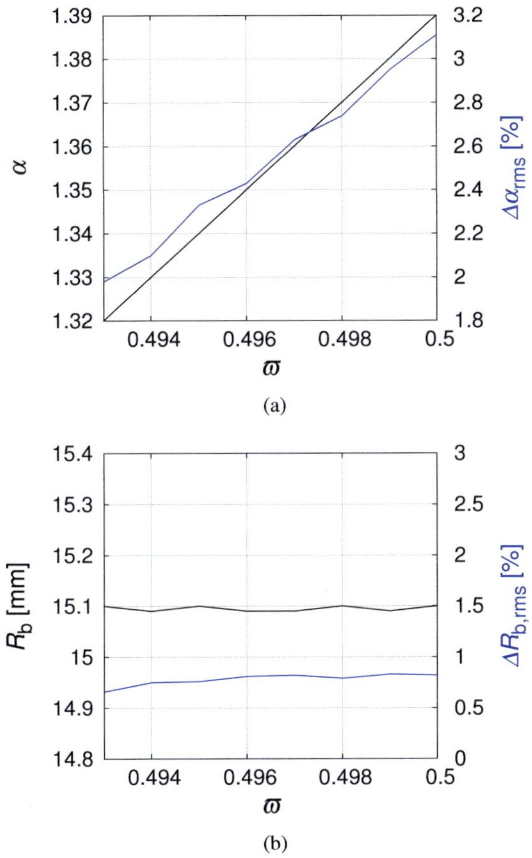

(a)

(b)

Fig. 3.7.: Influence of the magnetic field distribution at the emitter on α (a) and R_b (b) for the diode-MIG design ($\nu = 0.5$ and other parameters in Tab. 3.2)

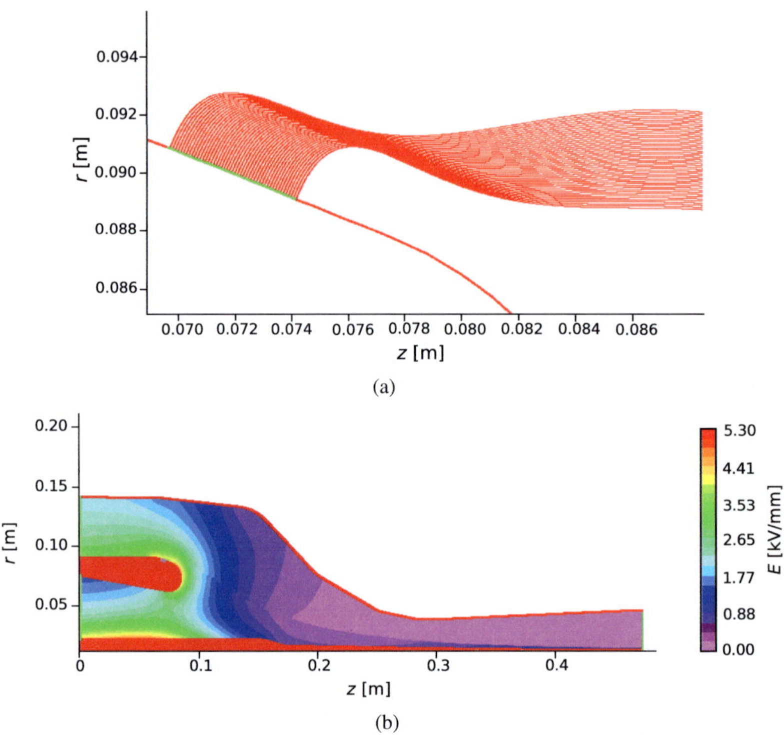

Fig. 3.8.: Quasi-laminar electron flow (a) and electric field strength (b) for the diode-type gun

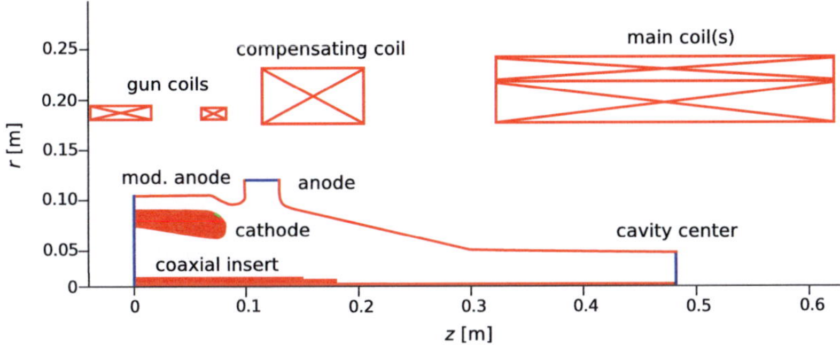

Fig. 3.9.: Geometry of the triode-type MIG and its magnet system

3.6. Design of a triode-type electron gun

The layout of the triode-type MIG and the magnet system is depicted in Fig. 3.9. The important design parameters are listed in Tab. 3.2, and the beam parameters of this triode-design (see Tab. 3.3) are obtained for a velocity ratio of approximately $\alpha = 1.35$, slightly higher than the nominal value (compared to Tab. 3.1). For this choice of α, higher velocity spreads are attained. If the beam parameters fulfill the specifications at this value, it can be expected that they are within specifications for other (lower) choices of the velocity ratio.

The beam radius R_b, its full radial width ΔR_b as well as the velocities and their spreads are well within specifications. The maximum deviations of the transverse velocity $\Delta\beta_{\perp,\mathrm{max}} = 1.33\%$ are widely below the limit as demanded in Eq. (3.13). In comparison to the diode design, as shown before, the triode-type MIG delivers are marginally lower beam quality, which can be assumed to be insignificant for reliable operation.

The velocity ratio α of the triode-type MIG can be adjusted smoothly by changing the applied voltage of the modulation anode V_mod without noticeably affecting other beam parameters, as predicted by adiabatic theory. Fig. 3.10(a) and 3.10(b) show the dependency of α and R_b on V_mod. α can be adjusted in a wide interval between 1.28 and 1.42. Its spread $\Delta\alpha_\mathrm{rms}$ increases corresponding to α, but is well within the acceptable limit $< 5\%$ for the complete desired operation regime. The beam radius R_b and ΔR_b do not change noticeably for tuning the beam parameters with V_mod. Necessary systematic changes for R_b can be achieved by adjusting the magnetic field at the emitter as it is utilized for diode-type electron guns.

The electron trajectories near the emitter surface are of the boundary flow type for $\varpi = 0.5$ and $\nu = 0.5$, as illustrated in Fig. 3.11(a). The electron trajectories are gathered to a small beam waist shortly after the emitter. As before for the diode design, the conditions in this region can be smoothly tuned by the two gun coil currents.

A sufficient margin to the maximum admissible electric field is maintained as plotted in Fig. 3.11(b). Between cathode and coaxial insert, the full accelerating voltage must be considered for their spacing ($V_\mathrm{acc} \approx 132\,\mathrm{kV} \rightarrow E_\mathrm{max} \approx 6.1\,\mathrm{kV/mm}$). To reduce the electric field in this region, the cathode's bottom thickness is reduced towards higher values in radius. The voltage drop between the cathode and modulation anode only amounts to approximately half the accelerating voltage, so the admissible electric field amplitude is higher in this region and the spacing of the electrodes can be closer ($V \approx 68\,\mathrm{kV} \rightarrow E_\mathrm{max} \approx 11.6\,\mathrm{kV/mm}$). The maximum electric field value occurs at the cathode region with a value of approximately $6.46\,\mathrm{kV/mm}$. In summary, the triode's modulation anode lowers the full value of the applied accelerating voltage along the electron trajectories. The technical constraint for the highest admissible electric field given in Eq. (3.19) can be fulfilled within a more compact triode design in comparison to the diode-type gun. The direct comparison of the radial space requirements for the diode- and triode type layouts can be found in Fig. 3.12. As a consequence, for the triode-type gun a far smaller bore-hole of the superconducting magnet is necessary. The additional degree of freedom for tuning the velocity ratio with V_mod makes the triode-type MIG preferable for a 4 MW coaxial-cavity gyrotron.

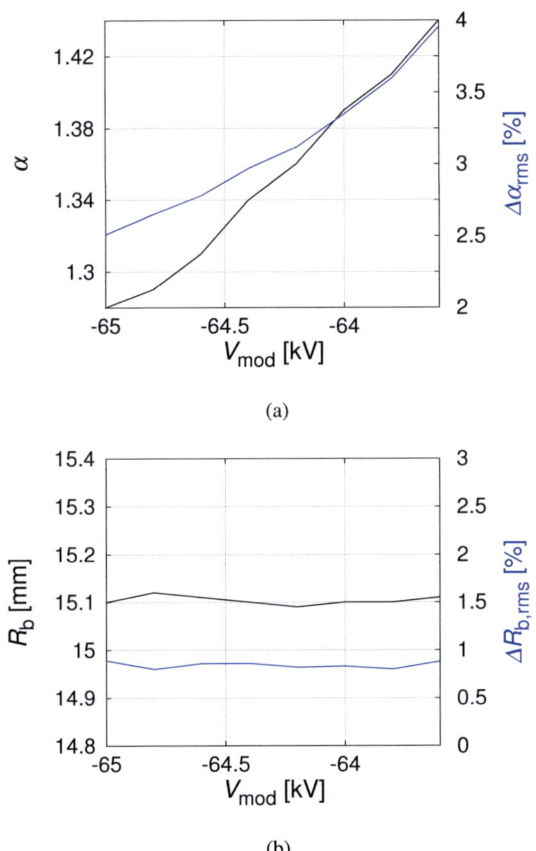

Fig. 3.10.: Influence of the triode's modulation voltage V_{mod} on α (a) and R_{b} (b), for parameters see Tab. 3.2

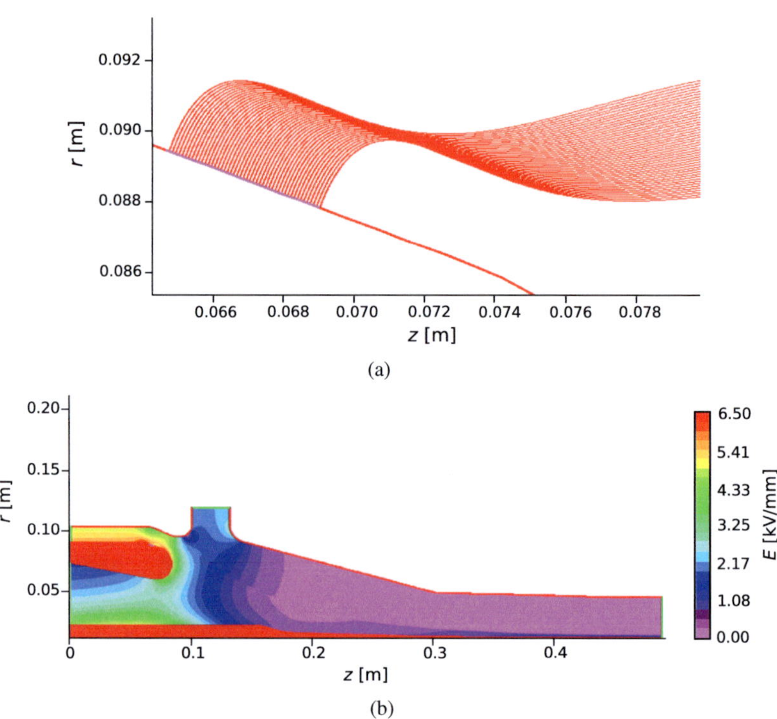

Fig. 3.11.: Quasi-laminar electron flow condition (a) and electric field distribution (b) for the triode-type electron gun ($\varpi = 0.5$ and $\nu = 0.5$)

		Diode	Triode
Average radius of the emitter [mm]	R_E	89.9	88.4
Emitter slant angle [°]	φ_C	21.9	20.4
Axial magnetic field at the emitter [mT]	$B_{\mathrm{E},r=0}$	259.2	253.5
Magn. field at the center of the emitter [mT]	B_E	157.3	179.4
Axial distance emitter and cavity [mm]	$\Delta z_{\mathrm{E-R}}$	410.8	416.3
Max. electric field at the insert [kV/mm]	$E_{\mathrm{I,max}}$	5.3	5.0
Max. electric field at the cathode [kV/mm]	$E_{\mathrm{C,max}}$	5.1	6.5
Electric field at the emitter surface [kV/mm]	E_E	3.8	5.6
Min. cathode-insert spacing [mm]	$\Delta r_{\mathrm{C-I}}$	42.8	41.6
Min. cathode-(mod.) anode spacing [mm]	$\Delta r_{\mathrm{C-A}}$	52.8	15.9
Radial offset of the (mod.) anode [mm]	R_A	145.4	104.0
Number of turns / I for gun coil 1 [A]	$N_{\mathrm{GC1}}/I_{\mathrm{GC1}}$	1139/26.81	1065/15.87
Number of turns / I for gun coil 2 [A]	$N_{\mathrm{GC2}}/I_{\mathrm{GC2}}$	1065/19.80	1065/27.83
Number of turns / I for comp. coil [A]	$N_{\mathrm{CC}}/I_{\mathrm{CC}}$	6139/−57.5	6139/−64.50
Number of turns / I for main coil 1 [A]	$N_{\mathrm{MC1}}/I_{\mathrm{MC1}}$	17275/111.16	17275/114.06
Number of turns / I for main coil 2 [A]	$N_{\mathrm{MC2}}/I_{\mathrm{MC2}}$	9333/111.16	9333/114.06
Number of turns / I for launcher coil [A]	$N_{\mathrm{LC}}/I_{\mathrm{LC}}$	6000/60	6000/60

Tab. 3.2.: Geometrical and magnetic parameters of the $TE_{-52,31}$ mode's diode- and triode-type electron gun designs

		Diode	Triode
Velocity ratio	α	1.39	1.35
Spread of the velocity ratio [%]	$\Delta\alpha_{\mathrm{rms}}$	3.11	2.97
Magnetic compression ratio	b	46.6	40.9
Average beam radius at the resonator [mm]	R_{b}	15.10	15.10
R.m.s. of radial beam width at resonator [%]	$\Delta R_{\mathrm{b,rms}}$	0.82	0.85
Total radial beam width at the resonator [mm]	ΔR_{b}	0.500	0.495
Relativistic Lorentz factor	γ	1.251	1.251
Normalized transverse velocity	$\beta_\perp$	0.490	0.483
R.m.s. of norm. transverse velocity spread [%]	$\Delta\beta_{\perp,\mathrm{rms}}$	1.14	1.16
Max. value of the norm. transverse velocity spread [%]	$\Delta\beta_{\perp,\mathrm{max}}$	1.32	1.33
Normalized axial velocity	$\beta_\parallel$	0.353	0.359
R.m.s. of norm. axial velocity spread [%]	$\Delta\beta_{\parallel,\mathrm{rms}}$	1.95	1.84
Max. value of norm. axial velocity spread [%]	$\Delta\beta_{\parallel,\mathrm{max}}$	2.55	2.79
Voltage depression in the cavity [kV]	ΔV	2.68	2.35
Voltage of the modulation anode [kV]	V_{mod}	n/a	−64.4

Tab. 3.3.: Optimized electron beam parameters for the $TE_{-52,31}$ mode's diode- and triode-type electron gun

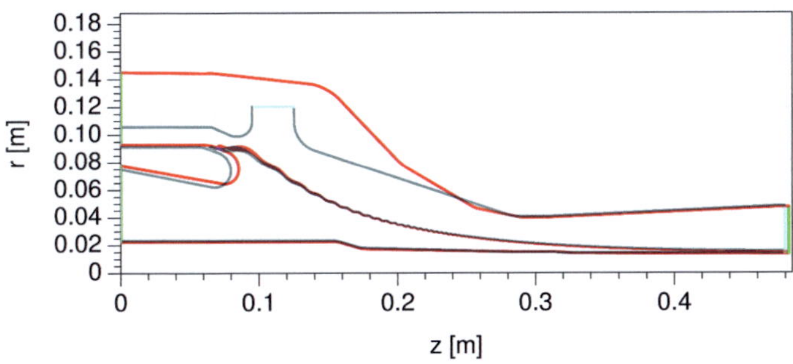

Fig. 3.12.: Comparison of the space requirements of diode- (red) and triode-type (green) MIG layouts

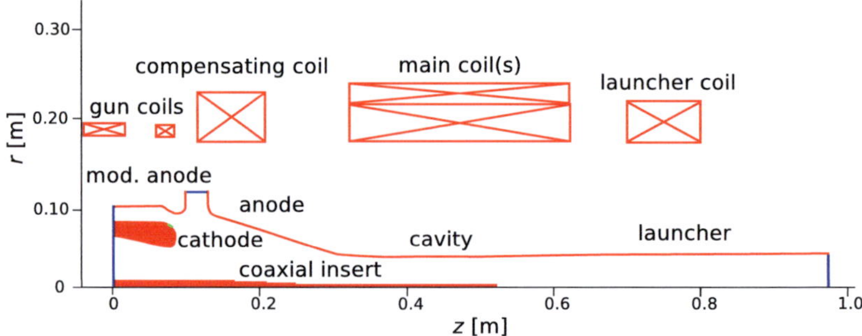

Fig. 3.13.: Complete triode layout including the additional launcher coil

In order to verify the alignment of the electron trajectories from the starting location at the emitter towards the launcher cuts, the complete geometry is modelled within the particle-in-cell simulation. Fig. 3.13 shows the gyrotron section starting from the triode-type electron gun to the end of the launcher cuts. The optimized uptaper after the cavity from Sec. 2.6 has an axial length of $87\,$mm. The quasi-optical output launcher as described in Ch. 4 has a length of $L_{\mathrm{L}} = 391\,$mm, resulting in a total axial dimension of $478\,$mm from the exit of the cylindrical cavity section to the end of the launcher cuts. In order to consider the diverging electron trajectories and to introduce a safety margin with respect to possible future design changes during the technical design phase, an additional magnetic coil close to the launcher (a so-called launcher coil) is introduced.

The additional launcher coil introduces a straightforward possibility to get influence on the alignment of the electron beam in the launcher region. This adjustment capability provides the advantage to prevent the beam to get too close to the launcher cuts and also provides the opportunity to reduce potential *after cavity interaction* (ACI) in this region. After the cavity, the magnetic field amplitude decreases and the electrons move in the same direction as the outgoing RF-field. The wave propagates in the uptapered waveguide where its cutoff frequency is decreasing along the axis. Therefore, it is possible for a certain cross-section, that the simultaneous tapering of the magnetic field and the waveguide radius wall result in the fulfillment of the electron-cyclotron-resonance-condition. To reduce the risk of possible after cavity effects, either the waveguide wall profile or the magnetic field (or both) have to be adjusted [SNJ10]. The latter can be easily done with an additional launcher coil.

The triode's setup is checked for regions where trapped electrons can occur, as described in Sec. 3.3 [PHA$^+$10]. The technical rear part of both design is simplified within the studies and thus no significant magnetic traps can be identified. For a technical layout of the electron gun and a consecutive introduction of more complex geometrical structures, the risk of generating a magnetic electron trap is present. This has to be considered in a further technical design phase.

3.7. Parameter studies for the diode- and triode-type electron gun

For the final electron gun designs, dependencies of beam parameters on the operating conditions are investigated to evaluate the performances and distinctions of both diode- and triode-type MIG layouts. The investigated characteristics are the space charge effects, the dependence of the velocity ratio α on the applied accelerating voltage and the sensitivity to mechanical parameter changes. Both diode- and triode-type designs are robust to restrictions, and especially the triode-type design shows several advantages.

3.7.1. Space charge effects

The impact of space charges on the beam properties (as outlined in Sec. 1.3.2) can be estimated by artificially reducing the beam current I_b of a design to zero. The electron charges constituting the beam lose their influence on the local electric field when they are decreased. When the current becomes zero, the electric field is defined only by the shape of the electrodes - the beam charges are not taken into account any more in the simulation. A design goal is to determine the properties of a beam

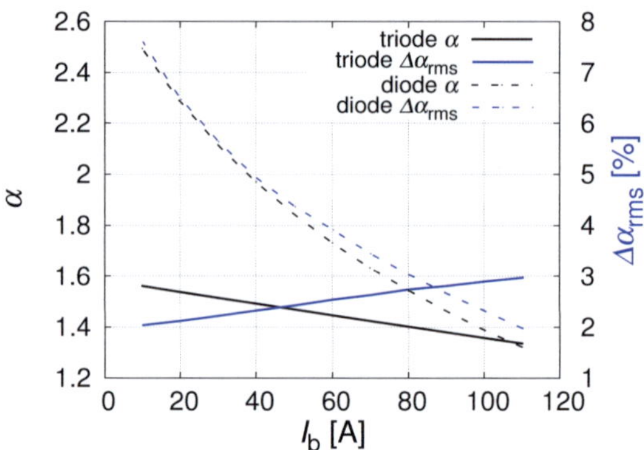

Fig. 3.14.: Dependence of the velocity ratio α on the beam current I_{b}

formed by a MIG as insensitive to the beam current as possible. The simulation results are given in Fig. 3.14. For the triode-type electron gun, the velocity ratio α is only marginally altered for currents lower than the nominal value, and this effect can be directly compensated by adjusting the modulation voltage V_{mod} as discussed before. However, the design of the diode-type MIG exhibits a much more distinct space-charge effect. The velocity ratio is increased to a regime where safe operation of the electron gun is rendered impossible. This effect is caused by the low electric field at the emitter surface of the diode-type MIG, which is in turn a direct consequence of the requirements on the maximum magnitude of the electric field, see Eq. (3.19). In contrast, within the design of the triode-type gun, the maximum electric field can be determined with a higher value because the voltage drop between cathode and modulation anode amounts to approximately half the accelerating voltage. The electric field is therefore defined mainly by the shape of the electrodes, and the beam charges have a lower influence in comparison to the diode-type design.

3.7.2. Dependency on the accelerating voltage

Fig. 3.15(a) and 3.15(b) show the dependence of the velocity ratio α and its spread $\Delta\alpha_{\mathrm{rms}}$ on the accelerating voltage V_{acc} for both electron gun designs. From adiabatic theory, a linear increase can be expected. This assumption is only roughly satisfied, indicating an influence of non-adiabatic effects. The triode-type electron gun shows a steeper incline of the velocity ratio as a consequence of the lower distinct impact of the space charges.

For a more accurate description of the MIGs' performance during start-up, the dependency of the beam current I_{b} respectively the current density at the emitter surface j_{E} on the accelerating voltage V_{acc} must be considered in the calculation. The basic correlation between the emitter temperature T and its current density j_{e} is given by the *Richardson-Dushman* equation:

$$j_{\mathrm{E}} = A_{\mathrm{G}} T^2 e^{\frac{-W}{k_{\mathrm{B}} T}} \tag{3.23}$$

j_{E} defines the current density at the emitter, T is the thermodynamic temperature of the material, W the work function of the metal and k_{B} the Boltzmann constant. The factor A_{G} is:

$$A_{\mathrm{G}} = \lambda_{\mathrm{R}} \cdot \frac{4\pi m_{\mathrm{e}} k_{\mathrm{B}}^2 e}{h^3} \tag{3.24}$$

λ_{R} is a material-specific correction factor that is typically of order 0.5. m_{e} and $-e$ are the mass and charge of an electron and h is Planck's constant. Eq. (3.23) can be further modified to include a Schottky effect in order to express the thermionic emission of a conductor caused by an electric field at its surface [Jr.87].

3.7.3. Sensitivity to mechanical errors

Within the optimization process of the diode- and triode-type electron guns in Sec. 3.5 and Sec. 3.6, usually a step-width of 0.1 mm has been utilized for the geometrical parameters. This value equates roughly the maximum permissible mechanical error in the fabrication of the MIGs. Studies are performed to identify the regions within the guns which are sensitive to parameter deviations and to estimate the corresponding variation of the beam

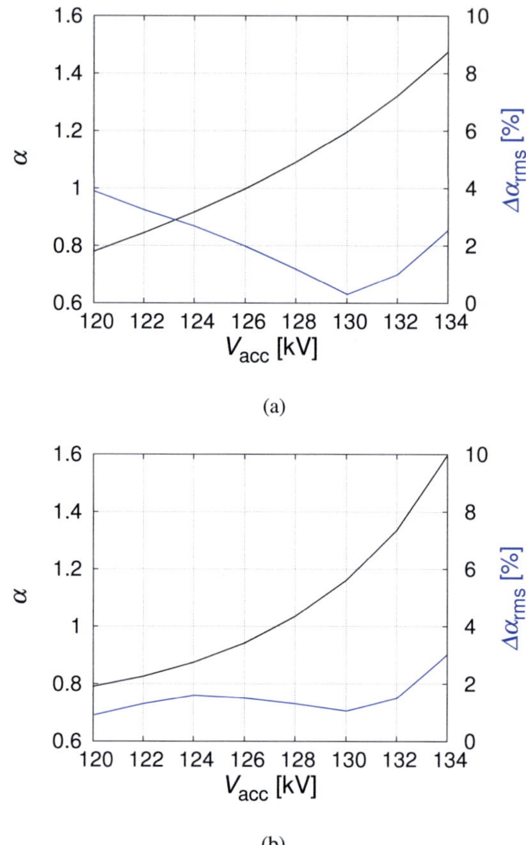

Fig. 3.15.: Influence of V_{acc} on the velocity ratio α for the diode- (a) and a triode-type (b) MIG layout

quality. For each geometrical parameter, modifications of three times the allowable mechanical tolerances (arbitrary equal $0.3\,\mathrm{mm}$) are introduced to the geometry of each design. As a result, three distinct regions can be identified for the diode- and triode-designs as marked in Fig. 3.16.

The blue regions in the vicinity of both electron guns' emitters and the triode's modulation anode are the most sensitive areas. The beam radius R_b is changed up to $0.05\,\mathrm{mm}$ and the velocity ratio α up to 0.07. The errors introduced to the cathode angle φ_C, to the radial position of the cathode and the radial alignment of the modulation anode have the largest impact on the beam properties. All beam deviations can be externally corrected by adjusting the currents applied to the gun coils and the modulating voltage. The velocity spread $\Delta\alpha_\mathrm{rms}$ will, however, be slightly increased by approximately $0.3\text{–}1.4\,\%$.

The changes to the parameters in the orange regions have a moderate influence on the beam parameters. The beam radius R_b is altered by only up to $0.02\,\mathrm{mm}$, whereas the velocity ratio α is shifted by up to 0.03. This effects can be compensated with almost no degradation of the velocity spread.

The geometrical parameters in the remaining regions only have a negligibly small influence on the beam. Its properties are only marginally altered when errors in this section occur.

The possibility to sensitively adjust the currents applied to the gun coils and the modulation voltage will provide options for compensating even multiple mechanical insufficiencies within the designs. It can therefore be said that tolerances of $\pm 0.3\,\mathrm{mm}$ are acceptable for both MIG designs.

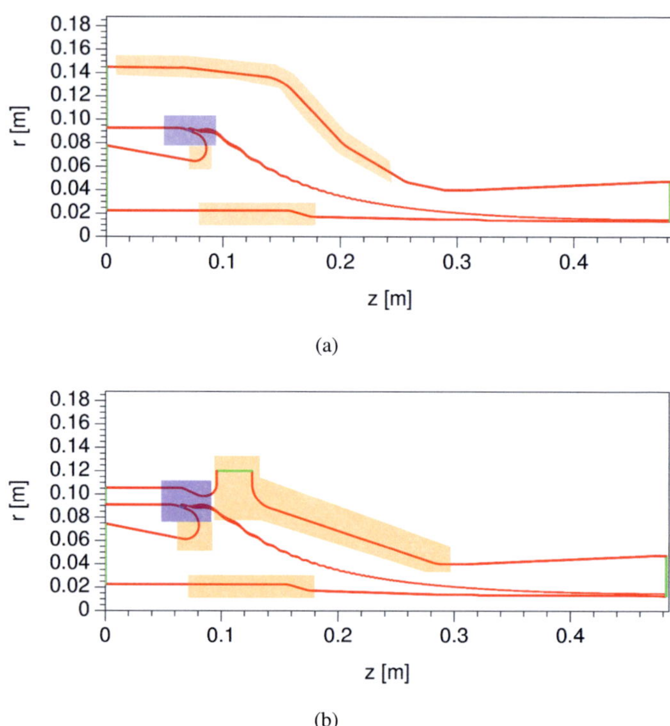

(a)

(b)

Fig. 3.16.: Regions of different sensitivity for mechanical tolerances (blue: high influence, orange: medium influence)

4. Quasi-optical output system

Quasi-optical mode converters in gyrotrons transform the high-order cavity mode into one or more fundamental Gaussian field distributions and separate the RF-beams from the spent electrons. The RF-beams exit the tube through CVD (chemical vapor deposition) diamond windows, which act as vacuum barriers. Technically available synthetic diamond windows can handle microwave beams up to $2\,MW$ [TAH$^+$01]. For a gyrotron with $4\,MW$ of generated RF-power, the waveguide mode has to be transformed into two Gaussian beams, which can be transmitted through two output windows. The output system consists of a so-called launcher, which acts as an antenna and several mirrors. In the launcher, which is in principle a hollow waveguide with one or two helical cuts at its end, a mode mixture is generated using specific inner wall surface perturbations [BD04]. This mode mixture is radiated as one or more fundamental Gaussian beams ($TEM_{0,0}$). The further conversion of the field is achieved by utilization of several reflectors (mirrors) to transform and guide the beams.
In this chapter, the design of a two-beam waveguide launcher using two cuts at its end is described. After the optimization of the launcher's surface structure, several simplification methods are discussed using two-dimensional filter functions. The design and filter techniques are verified using full wave calculation approaches.

4.1. Theoretical basics and calculation techniques

For the calculation of the electromagnetic field distribution inside the launcher it is reasonable to consider Maxwell's equations for the current and charge free ($\vec{j} = 0$, $\rho = 0$) case. Furthermore the field is assumed to be time harmonic ($e^{j\omega t}$) and the permittivity ($\varepsilon = \varepsilon_0$) and permeability ($\mu = \mu_0$) in vacuum are isotropic and homogeneous.

$$\nabla \times \vec{E} = -j\omega\mu_{\mathrm{o}}\vec{H} \quad \nabla \times \vec{H} = j\omega\varepsilon_0\vec{E} \tag{4.1}$$

$$\nabla \cdot \vec{E} = 0 \quad \nabla \cdot \vec{H} = 0 \tag{4.2}$$

Combining Eq. (4.1) and (4.2), an uncoupled second-order partial differential equation can then be written as [Bal38]:

$$\Delta U = \frac{-\omega^2}{\mu_0\varepsilon_0}U \tag{4.3}$$

$U(r, \varphi, z)$ is a scalar function that represents the vector potential. Eq. (4.3) can be identified as the *Helmholtz equation* using $k_0^2 = \omega^2/(\mu_0\varepsilon_0)$. Δ denotes the *Laplace operator* which is a differential operator with $\Delta = \nabla \cdot \nabla = \nabla^2$ and Eq. (4.3) expressed in cylindrical coordinates results in:

$$\frac{1}{r}\frac{\partial}{\partial r}\left(r\frac{\partial U}{\partial r}\right) + \frac{1}{r^2}\frac{\partial^2 U}{\partial\varphi^2} + \frac{\partial^2 U}{\partial z^2} = \frac{-\omega}{\mu_0\varepsilon_0}U \tag{4.4}$$

This equation can be solved using separated functions $P(r)$, $Q(\varphi)$ and $Z(z)$ which depend only on one space coordinate.

$$U(r, \varphi, z) = P(r)Q(\varphi)Z(z) \tag{4.5}$$

From Eq. (4.4) and Eq. (4.5) three separate differential equations with independent solutions can be obtained.

$$r^2\frac{d^2 P(r)}{dr^2} + r\frac{dP(r)}{dr} + \left[(k_{\mathrm{r}}r)^2 - m^2\right]P(r) = 0 \tag{4.6}$$

with

$$P(r) = A_1 J_{\mathrm{m}}(k_{\mathrm{r}}r) + B_1 N_{\mathrm{m}}(k_{\mathrm{r}}r) \tag{4.7}$$

$$\frac{d^2 Q(\varphi)}{d\varphi^2} + m^2 Q = 0 \tag{4.8}$$

with

$$Q(\varphi) = A_2 e^{-jm\varphi} + B_2 e^{+jm\varphi} \tag{4.9}$$

$$\frac{d^2 Z(z)}{dz^2} + k_z^2 Z(z) = 0 \tag{4.10}$$

with

$$Z(z) = A_3 e^{-jk_z z} + B_3 e^{+jk_z z} \tag{4.11}$$

$J_{\mathrm{m}}(k_{\mathrm{r}} r)$ are the Bessel and $N_{\mathrm{m}}(k_{\mathrm{r}} r)$ the Neumann functions. The constants $A_{1,2,3}$, $B_{1,2,3}$, m, k_{r} and k_{z} can be found using several boundary conditions.

For an ideal conducting surface at the waveguide wall ($r = R$) the following simplifications can be assumed:

- $E_\varphi(r = R, \varphi, z) = 0$ at the ideal conducting surface.

- The fields must be finite everywhere.

- The fields are periodic in φ and must repeat every 2π.

The Neumann function N_{m} has a singularity for $r \to 0$ and thus $B_1 = 0$ has to be chosen. In addition, according to the periodicity in φ the constant $m = 0, 1, 2, 3, \ldots$.

For waves that propagate only in $+z$-direction ($A_3 = 1$ and $B_3 = 0$) Eq. (4.5) reduces to:

$$U(r, \varphi, z) = A J_{\mathrm{m}}(k_{\mathrm{r}} r) \left[A_2 e^{-jm\varphi} + B_2 e^{+jm\varphi} \right] e^{-k_z z} \tag{4.12}$$

Considering modes, which only rotate in one direction with $A_2 = 1$ and $B_2 = 0$:

$$U(r, \varphi, z) = A J_{\mathrm{m}}(k_{\mathrm{r}} r) e^{-jm\varphi} e^{-jk_{\mathrm{z}} z} \tag{4.13}$$

The several field components for TE-modes ($E_{\mathrm{z}} = 0$) in a cylindrical coordinate system are consequently given as [Bal38]:

$$E_{\mathrm{r}} = -\frac{1}{\varepsilon_0 r} \frac{\partial U}{\partial \varphi} \tag{4.14}$$

$$E_{\varphi} = \frac{1}{\varepsilon_0} \frac{\partial U}{\partial r} \tag{4.15}$$

$$H_{\mathrm{r}} = -j \frac{1}{\omega \mu_0 \varepsilon_0} \frac{\partial^2 U}{\partial r \partial z} \tag{4.16}$$

$$H_{\varphi} = -j \frac{1}{\omega \mu_0 \varepsilon_0} \frac{1}{r} \frac{\partial^2 U}{\partial \varphi \partial z} \tag{4.17}$$

$$H_{\mathrm{z}} = -j \frac{1}{\omega \mu_0 \varepsilon_0} \left(\frac{\partial^2}{\partial z^2} + k_0^2 \right) U \tag{4.18}$$

With the use of Eq. (4.13) and (4.15), the electric field component of E_{φ} can be written as:

$$E_{\varphi}(r, \varphi, z) = \frac{1}{\varepsilon_0} \frac{\partial U}{\partial r} = k_{\mathrm{r}} \frac{A}{\varepsilon_0} J'_{\mathrm{m}}(k_{\mathrm{r}} r) e^{-jm\varphi} e^{-jk_{\mathrm{z}} z} \tag{4.19}$$

where

$$' = \frac{\partial}{\partial (k_{\mathrm{r}} r)} \tag{4.20}$$

To fulfill the boundary condition on the waveguide wall $E_{\varphi}(r = R, \varphi, z) = 0$ for TE-modes, $J'_{\mathrm{m}}(k_{\mathrm{r}} R) = 0$ has to be chosen accordingly.

$$J'_{\mathrm{m}}(k_{\mathrm{r}} r) = 0 \rightarrow k_{\mathrm{r}} R = \chi_{\mathrm{m,p}} \rightarrow k_{\mathrm{r}} = \frac{\chi_{\mathrm{m,p}}}{R} \tag{4.21}$$

$\chi_{\mathrm{m,p}}$ represents the p-th zero ($p = 1, 2, 3, \ldots$) of the derivative of the Bessel function J_{m} of the first kind with the order m ($m = 1, 2, 3, \ldots$). In general, $k_0 = \sqrt{k_{\mathrm{z}}^2 + k_{\mathrm{r}}^2}$ and cutoff (with index $\perp$) is defined when $k_{\mathrm{z}} = 0$.

$$k_\perp = \omega_\perp \sqrt{\mu_0 \varepsilon_0} = 2\pi f_\perp \sqrt{\mu_0 \varepsilon_0} = \frac{\chi_{\mathrm{m,p}}}{R} \tag{4.22}$$

For a detailed characterization of the launcher attributes it is advantageous to describe the waveguide modes as a bunch of rays. These rays are tangential to a circle with the caustic radius R_{c} and reflected on the launcher surface as shown in Fig. 4.1.

The field can be split into an incident and a reflected wave with the amplitude A.

$$u = u^{(in)} + u^{(out)} = A J_{\mathrm{m}}(k_{\mathrm{r}} r) e^{-jm\varphi} e^{-jk_{\mathrm{z}} z} \tag{4.23}$$

The Bessel function J_{m} at any position x can be replaced by the sum of two Hankel functions H_{m}. The Hankel function of the first kind represents the incident and the Hankel function of the second kind the reflected wave respectively.

$$J_{\mathrm{m}}(x) = \frac{1}{2}(H_{\mathrm{m}}^{(1)}(x) + H_{\mathrm{m}}^{(2)}(x)) \tag{4.24}$$

with

$$H_{\mathrm{m}}^{(1)}(x) = J_{\mathrm{m}}(x) + j N_{\mathrm{m}}(x) \tag{4.25}$$

and

$$H_{\mathrm{m}}^{(2)}(x) = J_{\mathrm{m}}(x) - j N_{\mathrm{m}}(x) \tag{4.26}$$

In general the two Hankel functions can be transformed into each other by complex conjugation.

$$H_{\mathrm{m}}^{(1)}(x) = \left(H_{\mathrm{m}}^{(2)}(x) \right)^* \tag{4.27}$$

An approximation for $H_{\mathrm{m}}^{(2)}(z)$ for $z > m$ is given in [MF53]:

$$H_{\mathrm{m}}^{(2)}(z) \approx \sqrt{\frac{2}{\pi\sqrt{z^2 - m^2}}}\, e^{j\left(-\sqrt{z^2 - m^2} + m\arccos\left(\frac{m}{z}\right) + \frac{\pi}{4}\right)} \qquad (4.28)$$

As a consequence the phase of the reflected (outgoing) wave is:

$$arg(u^{(out)}(r, \varphi, z)) \approx -\sqrt{k_{\mathrm{r}}^2 r^2 - m^2} + m\arccos\left(\frac{m}{k_{\mathrm{r}} r}\right) + \frac{\pi}{4} - m\varphi - k_{\mathrm{z}} z \qquad (4.29)$$

The vector $\vec{N}$ normal to the planes of constant phase a for the reflected wave represents the direction of each bunch of rays. A solution for $\vec{N}$ can be obtained using an implicit equation $a - arg(u^{(out)}) = 0$.

$$\vec{N}(r, \varphi, z) = \nabla\left(a - arg\left(\frac{A}{2} H_{\mathrm{m}}^{(2)} e^{-jm\varphi} e^{-jk_{\mathrm{z}} z}\right)\right) \qquad (4.30)$$

Summaries of the related theory can be found in [Mic98] and [Dru02]. Detailed descriptions and derivations are given in [Wei69] and [Sil86].
With the modes' ray representation several geometrical parameters can be derived. The *spread angle* 2ψ is defined as the angle between two reflection points.

$$2 \cdot \psi = 2 \cdot \arccos\left(\frac{m}{\chi_{\mathrm{m,p}}}\right) \qquad (4.31)$$

The launcher has a total length L_{L} and a radius R_{L}. The ratio between the launcher radius at its input $R_{\mathrm{L,0}}$ and the cavity radius R_{R} is called the *oversize factor*. The length of the launcher depends strongly on the chosen oversize factor but has an upper limit due to the radial widening of the electron beam. The distance between two reflection points on the waveguide walls along the waveguide axis can be described by the *Brillouin length* L_{B}.

$$L_{\mathrm{B}} = \frac{2R_{\mathrm{L}}^2}{\chi_{\mathrm{m,p}}} \sqrt{\left(1 - \left(\frac{m}{\chi_{\mathrm{m,p}}}\right)^2\right)\left(k_0^2 - \left(\frac{\chi_{\mathrm{m,p}}}{R_{\mathrm{L}}}\right)^2\right)} \qquad (4.32)$$

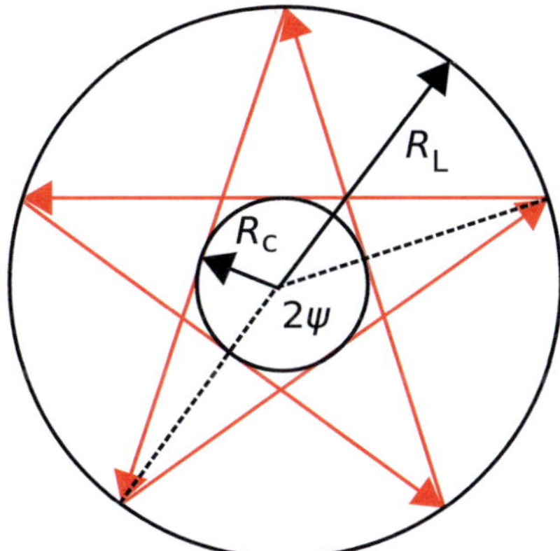

Fig. 4.1.: The spread angle 2ψ between two reflection points

The so-called *Brillouin angle* θ is given as:

$$\theta = \arcsin\left(\frac{k_\mathrm{r}}{k_0}\right) = \arcsin\left(\frac{\chi_{\mathrm{m,p}}}{R_\mathrm{L} \cdot k_0}\right) \tag{4.33}$$

4.2. Utilized analysis and optimization routine

The utilization of a high-order operating mode results in a highly oversized waveguide with large diameter. The ratio of the launcher diameter to the wavelength is typically significantly high, as a consequence, the field inside the launcher can be described by means of the quasi-optical method as introduced in Sec. 4.1. The synthesis of the surface is based on the idea to create a mode mixture, which has a Gaussian-like field distribution [BD04]. The internal field remains paraxial for launchers with small surface perturbations and a smooth contour. The numerical method applied to design the two-beam waveguide launcher is described in [JTP+09] and the following equations summarize the core issues. The *scalar diffraction*

equation Eq. (4.36) is utilized to analyze the paraxial wave fields.
The contour surface of the untapered launcher $R_{\mathrm{L}}(\varphi, z)$ is modeled as:

$$R_{\mathrm{L}}(\varphi, z) = R_{\mathrm{L},0} + \Delta R_{\mathrm{L}}(\varphi, z) = R_{\mathrm{L},0} + \sum_{\mathrm{a,b}} \Delta R_{\mathrm{L,ab}} \delta(\varphi - \varphi_{\mathrm{a}}, z - z_{\mathrm{b}})$$

(4.34)

with

$$\delta(\varphi - \varphi_{\mathrm{a}}, z - z_{\mathrm{b}}) = \begin{cases} 1, & \varphi = \varphi_{\mathrm{a}} \text{ and } z = z_{\mathrm{b}} \\ 0, & \varphi \neq \varphi_{\mathrm{a}} \text{ or } z \neq z_{\mathrm{b}} \end{cases}$$

(4.35)

$R_{\mathrm{L},0}$ is the launcher's average radius and ΔR_{L} depicts the surface perturbations. In general, if the deformations are small ($\Delta R_{\mathrm{L}} \ll R_{\mathrm{L},0}$) and relatively smooth ($\nabla |R_{\mathrm{L}}(\varphi, z)| \ll 1$), the field on the closed launcher surface S can be determined in terms of the scalar diffraction equation.

$$u(\vec{x}) = -\int\int \left(u_{\mathrm{c}}(\vec{x_0})\nabla g(\vec{x} - \vec{x_0}) - g(\vec{x} - \vec{x_0})\nabla u_{\mathrm{c}}(\vec{x_0}) \right) \vec{n_0} ds_0$$

(4.36)

The fields on the perturbated u_{c} surface can be described as the field u on the unperturbated wall with an additional phase corrector.

$$u_{\mathrm{c}}(\vec{x}) = u(\vec{x}) e^{\mp j 2 k_0 \Delta R_{\mathrm{L}} \cos(\alpha)}$$

(4.37)

with

$$|\vec{x} - \vec{x_0}| = \sqrt{(z - z_0)^2 + 4R_{\mathrm{L}}^2 \sin^2 \left(\frac{\varphi - \varphi_0}{2} \right)}$$

(4.38)

and

$$\cos(\alpha) = \frac{\sqrt{\chi_{\mathrm{m,p}}^2 - m^2}}{k_0 \cdot R_{\mathrm{L}}}$$

(4.39)

g is the Green's function in free space which is used as an integral kernel to solve the scalar diffraction equation.

$$g(\vec{x} - \vec{x_0}) = \frac{e^{\mp(jk_0|\vec{x}-\vec{x_0}|)}}{4\pi|\vec{x} - \vec{x_0}|}$$

(4.40)

The unit vector along the direction normal to the launcher's surface is given as $\vec{n_0}$. Assuming the coupling of TE- to TM-modes to be very weak and negligible [JTP$^+$09], Eq. (4.36) can be simplified to:

$$u_{i+1}(\vec{x}) = \int_{-\infty}^{L_L} \int_0^{2\pi} u_i(\vec{x_0})e^{\mp j2k\Delta R_L(\varphi_0,z_0)\cos(\alpha)}$$
$$\cdot \frac{\partial g(\vec{x} - \vec{x_0})}{\partial n_0} R_{L,0}d\varphi_0 dz_0$$

(4.41)

u_{i+1} denotes the new field distribution on the deformed launcher wall. The upper limit L_L of the first integral describes the length of the launcher, $i = 0 \ldots I$ is an iterative number and the directions $-$ and $+$ describe the forward and backward propagation of the field inside the launcher. The number I has to be equal or larger than L_L/L_B, so that the scattering field can cover the entire launcher surface. The integral region with respect to z, starting from minus infinity, has to be shortened to achieve sufficiently exact numerical results. Comparing the fields u_i and u_{i+1}, the integration results in the propagation of the field by one Brillouin length L_B in z-direction.

Eq. (4.36) and (4.41) are actually convolution integrals and can be evaluated by means of the two-dimensional fast Fourier transformation (FFT).

4.3. Design results for a two-beam launcher

For the evaluation of the launcher's output beams usually the *vector correlation coefficient* η with respect to the fundamental Gaussian distribution is utilized as a measure of the beam quality. If the function u_2 is an ideal Gaussian distribution, η describes the Gaussian mode content of the field distribution u_1 [CDK$^+$06].

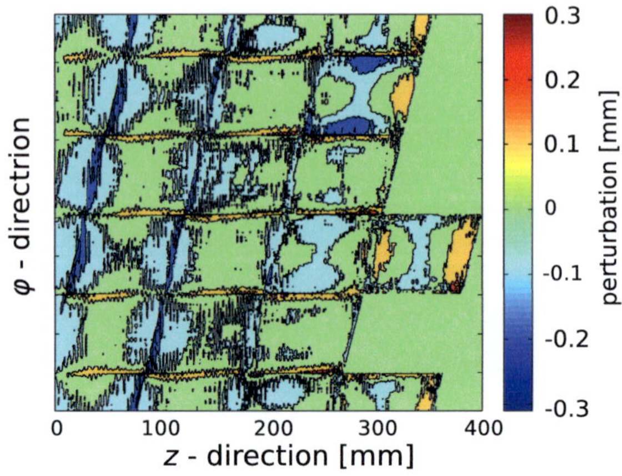

Fig. 4.2.: Optimized wall perturbation on launcher surface with two cuts (untapered)

$$\eta = \frac{|\int_S u_1^* u_2 ds|^2}{\int_S |u_1|^2 ds \cdot \int_S |u_2|^2 ds} \tag{4.42}$$

The symbol "*" represents the conjugated conversion of a complex function.

For the synthesis in [JTP$^+$09] an iterative algorithm is utilized and implemented in a software package *TWL_DO*. This code is used to achieve the optimized launcher surface perturbations as described in this work. The mathematical basics are given in Sec. 4.2. The main idea is that an ideal field distribution is injected during every optimization cycle from the output side into the launcher aperture. The surface wall structure is iteratively optimized to achieve the highest possible correlation coefficients of each outgoing wave beam with the desired distribution. In the case of the 4 MW two-beam launcher design, as shown in this work, the ideal RF-output consists of two beams with Gaussian field distribution.

The launcher surface profile including two cuts after optimization can be found in Fig. 4.2. The alignment of the perturbations on the surface form several areas (parallelograms) and on each of these so-called *Brillouin zones* with length L_B one beam is reflected. The resolution is set arbitrary to 1024 points in φ- and 1793 points in z-direction in order to achieve reasonable calculation time and sufficient accuracy. The two cuts have a length of 92.3 mm each. The radius of the launcher at its input is $R_\mathrm{L,0} = 50.633$ mm (for an oversize factor of 1.07) and the total length is $L_\mathrm{L} = 391$ mm. The surface perturbations have a maximum depth of -0.27 mm $< \Delta R_\mathrm{L} \leq 0.35$ mm. The minimum curvature radius which is necessary for manufacturing the surface structure is determined to be 3.82 mm.

Fig. 4.3(a) shows the z-component of the $\vec{H}$-field on the wall. The two output beams have a fundamental Gaussian mode content of $\eta_1 = 98.3\%$ and $\eta_2 = 96.9\%$ respectively. Each beam has approximately 10 reflections along the surface. A longer launcher would only marginally increase the beams' qualities. The corresponding phase distribution can be found in Fig. 4.3(b). The phase distribution shows a linear characteristic for each of the two output beams at the last reflection zone, which is characteristic for distributions with high fundamental Gaussian mode content.

For preventing wave reflection at some perturbed positions along the launcher surface, and thus preventing parasitic oscillations in the converter region by the spent electron beam, a positive slope taper τ is usually introduced on the wall radius of the launcher [TCP97]. As a consequence, Eq. (4.34) is supplemented to:

$$
\begin{aligned}
R_\mathrm{L}(\varphi, z) &= R_\mathrm{L,0} + \tau z + \Delta R_\mathrm{L}(\varphi, z) \\
&= R_\mathrm{L,0} + \tau z + \sum_{\mathrm{a,b}} \Delta R_\mathrm{L,ab}\delta(\varphi - \varphi_\mathrm{a}, z - z_\mathrm{b})
\end{aligned}
\qquad (4.43)
$$

In order to simplify the coupler's optimization process, the taper of the launcher is neglected ($\tau = 0$) within this first approach. Usually the introduction of a tapered launcher surface during optimization of the perturbations does not result in a lower Gaussian mode content of the output beams. $\tau > 0$ leads to a shift of each Brillouin zone towards higher z-values and thus in a slightly longer launcher layout. The upcoming possible conflict

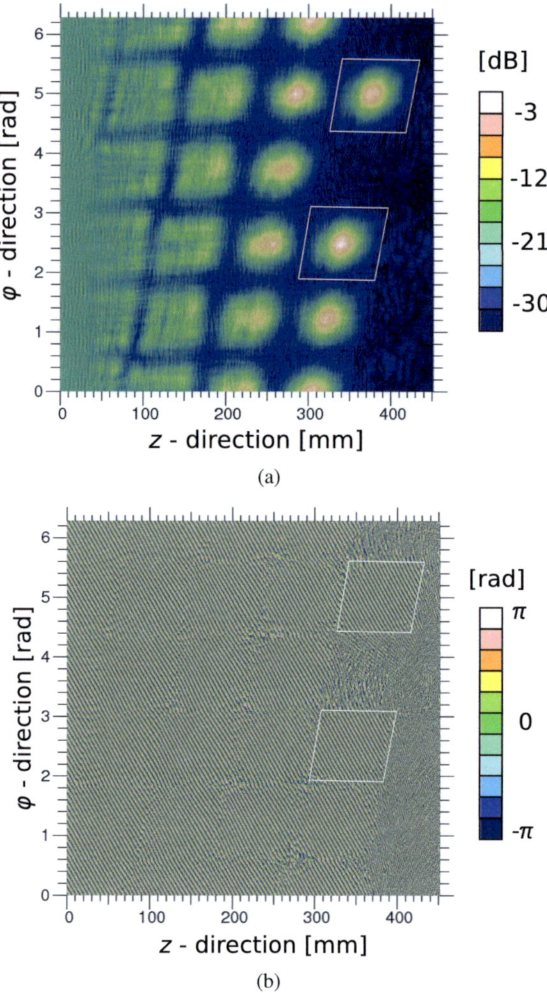

(a)

(b)

Fig. 4.3.: H_z-magnitude (a) and phase (b) on the optimized two-beam launcher wall as illustrated in Fig. 4.2

of the shifted cuts with the opening electron beam has to be carefully considered and avoided.

4.4. Analysis of small deviations on the launcher surface

The launcher surface, as introduced in Sec. 4.3, has a relatively complex surface structure in order to generate two high quality output beams. The optimized surface structure in Fig. 4.2 consists partially of distributed steep slopes and deep perturbations which complicate the manufacturing process. In the next sections, the surface structure is analyzed and possible smoothening methods are discussed to allow an economic machining process of the launcher. In a first step, an easy averaging of the surface perturbations is discussed and further two-dimensional filter techniques are introduced. For each simplification technique, the advantages and influences on the beam quality are discussed.

4.4.1. Smoothening techniques for the launcher surface

A fast approach for the simplification of the surface structure is a simple two-dimensional averaging of the perturbation depths. The mean value for one discrete point $\Delta R_{\mathrm{L},ij} = \Delta R_{\mathrm{L}}(i\Delta\varphi, j\Delta z)$ can be determined within a quadratic surface section with an odd number of points along one side. The single discrete values are given by $\Delta R_{\mathrm{L},ij}$ and their number is $M \times N$. $\Delta\varphi$ and Δz are the discretization steps in φ- and z-direction. The two-dimensional standard deviation σ and the variance $Var = \sigma^2$ can be utilized to characterize the perturbation depth. The average perturbation depth on the launcher surface can be determined by the mean value of $\Delta R_{\mathrm{L},ij}$.

$$\overline{\Delta R_{\mathrm{L}}} = \frac{1}{M \cdot N} \sum_{i=1}^{M} \sum_{j=1}^{N} \Delta R_{\mathrm{L},ij} \qquad (4.44)$$

The variance determines how far values are located from the mean and indicates the "complexity" of the structure. The variance is the square of the standard deviation σ.

$$\sigma = \left(\frac{1}{M \cdot N - 1} \sum_{i=1}^{M} \sum_{j=1}^{N} (\Delta R_{\mathrm{L,ij}} - \overline{\Delta R_{\mathrm{L}}})^2 \right)^{1/2} \tag{4.45}$$

For a more effective simplification of the launcher's surface structure, two-dimensional filtering techniques are utilized. The launcher perturbations can be described as a two-dimensional discrete signal. To achieve reasonable filter functions, the two-dimensional filter response is derived from a one-dimensional digital filter with the desired characteristics. In the following, the mathematical basics are introduced to derive the utilized two-dimensional filter functions.

In its simplest form for discrete systems, the transfer function $H(z)$ is the linear mapping of the $\mathcal{Z}$-*transform* of the input $X(z)$ to the output $Y(z)$ [KK09].

$$Y(z) = H(z) \cdot X(z) \tag{4.46}$$

The frequency response of a digital filter can be interpreted as the transfer function $H(z)$ evaluated at $z = e^{j\omega}$.

A given one-dimensional filter response can be converted into a two-dimensional response by the *frequency transformation method* [Lim90]. The frequency transformation method preserves most of the characteristics of the one-dimensional filter, particularly the transition bandwidth and ripple characteristics. This method uses a transformation matrix t, a set of elements that defines the frequency transformation. The corresponding two-dimensional continuous filter response can thus be derived by using the Fourier transform T of the transformation matrix t. Consider a zero-phase one-dimensional filter $h(n)$ with length $2N+1$ then the frequency response $H(\omega)$ can be expressed as:

$$H(\omega) = \sum_{n=-N}^{N} h(n)e^{-j\omega n} = \sum_{n=0}^{N} c(n)(\cos(\omega))^n \tag{4.47}$$

In Eq. (4.47), the sequence $c(n)$ is not the same as $h(n)$, but can be obtained from $h(n)$. The two-dimensional frequency response $H(\omega_1, \omega_2)$ is given by:

$$H(\omega_1, \omega_2) = H(\omega)|_{\cos(\omega)=T(\omega_1,\omega_2)} = \sum_{n=0}^{N} c(n) \left(T(\omega_1, \omega_2)\right)^n \quad (4.48)$$

where $T(\omega_1, \omega_2)$ is the Fourier transform of a finite-extent zero-phase sequence t.

$$T(\omega_1, \omega_2) = \sum_{n_2} \sum_{n_1} t(n_1, n_2) e^{-j\omega_1 n_1} e^{-j\omega_2 n_2} \quad (4.49)$$

An example of $T(\omega_1, \omega_2)$ often used in practice and known as the *McClellan transformation* is [MC77]:

$$T(\omega_1, \omega_2) = -\frac{1}{2} + \frac{1}{2} \cos(\omega_1) + \frac{1}{2} \cos(\omega_2) + \frac{1}{2} \cos(\omega_1) \cos(\omega_2) \quad (4.50)$$

The frequency response vector is the inverse Fourier transform of $H(\omega_1, \omega_2)$.

$$h(n_1, n_2) = \frac{1}{(2\pi)^2} \int_{-\pi}^{\pi} \int_{-\pi}^{\pi} H(\omega_1, \omega_2) e^{j\omega_1 n_1} e^{j\omega_2 n_2} d\omega_1 d\omega_2 \quad (4.51)$$

The basic idea behind the transformation method can be used to develop two-dimensional filters with desired properties. In the next section, such two-dimensional filters are utilized to simplify the structure of the launcher perturbations.

4.4.2. Averaging and low-pass filtering of launcher perturbations

As described above, a fast approach to smoothen the launcher's surface structure is the two-dimensional averaging of the discrete points after optimization. The values are calculated using the mean within a quadratic section with an odd number of points along one side. After each averaging step, the vector correlation coefficients of both beams and the variance of the surface perturbation are determined. The corresponding numbers can be found in Tab. 4.1. The applied averaging only marginally lowers the complexity of the surface perturbations, but most of sharp edges are suppressed. A simple averaging for every point with its surrounding neighbors does not lower the quality of the output beams.

	η_1 [%]	η_2 [%]	$\overline{\Delta R}$ [mm]	σ [mm]	σ^2 [mm^2]
w/o averaging	97.80	96.90	0.0128	0.0562	0.0032
avg. 3×3 pts.	98.28	96.86	0.0128	0.0558	0.0031
avg. 6×6 pts.	98.31	96.90	0.0128	0.0547	0.0030
avg. 9×9 pts.	98.39	96.96	0.0128	0.0535	0.0029

Tab. 4.1.: Vector correlation coefficients η_1 and η_2 of both beams after surface averaging

In a next approach, discrete two-dimensional low-pass filters are designed as described in Sec. 4.4.1 and applied to the optimized structure from Sec. 4.3. To achieve fast calculations *Chebyshev* filters of the second type are introduced. Chebyshev filters have a steeper slope between pass- and stopband as (for example) *Butterworth*-filters of the same order [KK09]. In addition, Chebyshev filters of type II have a flat characteristic in the passband and do not influence the necessary smooth perturbations of the launcher surface.

The one-dimensional frequency responses of a typical Chebyshev type II filter is shown in Fig. 4.4. The normalized sampling rate is ω_n, the passband frequency is set to $\omega_p = 0.3 \cdot \omega_n$ and the corresponding stopband frequency to $\omega_s = 0.4 \cdot \omega_n$. For illustration, this choice for the pass- and stopband delivers a filter order $m = 8$, which guarantees fast calculation. The difference of the perturbations before and after filtering is plotted in Fig. 4.5(a). The surface is mostly simplified at the outer area of each Brillouin zone where usually the perturbations with highest amplitude occur. The maximal difference of the perturbation after filtering is 0.08 mm, which is a

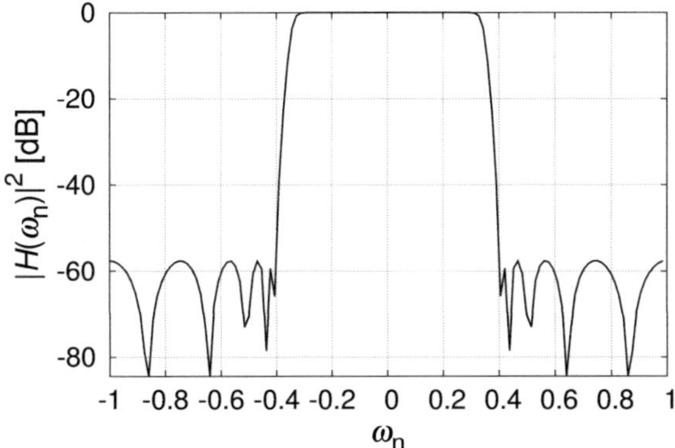

Fig. 4.4.: One-dimensional frequency response of a Chebyshev filter (type II)

strong simplification in comparison to the maximum perturbation depth of $-0.27\,\mathrm{mm} < \Delta R_\mathrm{L} < 0.35\,\mathrm{mm}$ for the original surface structure. A detailed surface section before and after filtering is plotted in Fig. 4.5(b) at an arbitrary position along the launcher surface at $z = 112\,\mathrm{mm}$ and $2.15 < \varphi\,(\mathrm{rad}) < 3.07$ (as indicated by the red arrow in Fig. 4.5(a)). The surface structure is strongly simplified due to the missing high frequency portions. Steep gradients along the surface are suppressed and deep insections are attenuated.

Until now the pass and stop frequency are chosen arbitrarily. To determine the frequency limit for filtering and to guarantee high Gaussian mode content of the output beams, the two-dimensional filter edge is shifted towards lower frequencies and the corresponding vector correlation coefficients η_1 and η_2 for both beams are determined. In Tab. 4.2, r_p defines the allowed ripples in the passband of the filter and r_s the attenuation in the stopband in dB. r_p is chosen to be small in order to keep the low frequency portions undisturbed. To achieve an odd-length filter order m for a faster transformation of the one- to the two-dimensional filter function, the attenuation r_s is arbitrarily chosen to be $> 50\,\mathrm{dB}$ [HB92].

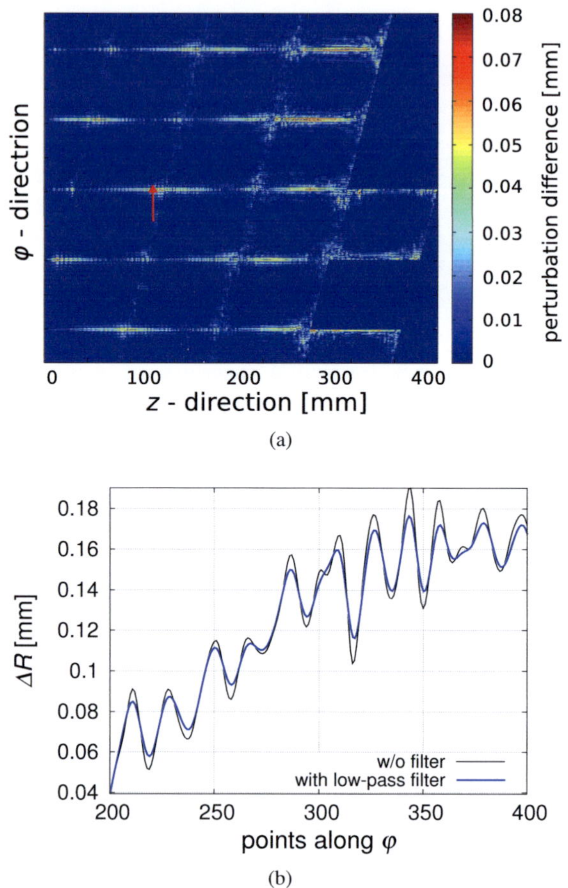

(a)

(b)

Fig. 4.5.: Difference between original and filtered perturbation depth (a) and detail along the red arrow (b) for $2.15 < \varphi < 3.07$ and $z = 112\,\text{mm}$

ω_p	ω_s	r_p	r_s	m	$\Delta R_\mathrm{L,min}$ [mm]	$\Delta R_\mathrm{L,max}$ [mm]	η_1 [%]	η_2 [%]
original					-0.3105	+0.3166	98.31	96.88
0.80	0.90	0.5	50	9	-0.3017	+0.3056	98.31	96.88
0.70	0.80	0.5	50	9	-0.2989	+0.3016	98.32	96.90
0.60	0.70	0.5	60	11	-0.2956	+0.2969	98.33	96.92
0.50	0.60	0.5	70	13	-0.2915	+0.2915	98.36	96.93
0.40	0.50	0.5	70	13	-0.2886	+0.2880	98.37	96.95
0.30	0.40	0.5	60	11	-0.2871	+0.2860	98.37	96.95
0.25	0.35	0.5	80	13	-0.2630	+0.2595	98.37	96.85
0.24	0.34	0.5	80	11	-0.2179	+0.2148	93.96	89.57
0.23	0.33	0.5	50	9	-0.1965	+0.1958	81.37	74.76

Tab. 4.2.: Vector correlation coefficients of both beams after surface averaging

The low-pass filtered surface structure delivers a high-quality beam even if the filter characteristic is chosen to be relatively small ($\omega_{\mathrm{s}} \approx 0.4 \cdot \omega_{\mathrm{n}}$). Starting from $\omega_{\mathrm{p}} < 0.25$, an even smaller filter response reduces the beam quality rapidly. Nevertheless, it is possible to simplify the launcher's surface structure noticeable without lowering the vector correlation coefficient of each beam.

4.5. Verification of the launcher design and introduced smoothening

The commercial full wave three-dimensional code *Surf3D* [Nei04] is utilized to calculate the field radiated from the launcher in order to verify the optimized surface perturbations as introduced in Sec. 4.3. The applied method is based on the *electric field integral equation* and explained in the following:

The incident tangential field $\vec{E}_{\mathrm{t}}^{i}$ induces a surface current density $\vec{j}_{\mathrm{S}}$ on a reflecting conducting surface S [Bal38].

$$\vec{E}_{\mathrm{t}}^{i}(r = R_{\mathrm{S}}) = j\eta k_0 \left[\oint_{\mathrm{S}} \left(1 + \frac{1}{k_0^2} \nabla \nabla' \right) \vec{j}_{\mathrm{S}}(r') g(R_{\mathrm{S}}, r') ds' \right] \quad (4.52)$$

The usual coordinates give the observing location and the coordinates denoted with a dash ($'$), which have to be integrated, are on the surface of the conductor. $g(r, r')$ is the free space Green's function with the distance $R = |r - r'|$ and $\eta = \sqrt{\mu_0/\varepsilon_0}$ [Roa82].

$$g(r, r') = \frac{e^{-jk_0 R}}{4\pi R} = \frac{e^{-jk_0|r-r'|}}{4\pi|r - r'|} \quad (4.53)$$

The tangential electric field E_{t} vanishes on the surface ($r = R_{\mathrm{s}}$) of the conductor and so the tangential electric field of the incident and outgoing (scattered) wave have to be equal but phase shifted.

$$E_{\mathrm{t}}^{i}(r = R_{\mathrm{s}}) = -E_{\mathrm{t}}^{s}(r = R_{\mathrm{s}}) \quad (4.54)$$

For every vector component, the unknown surface current j_S is discretized and the electric field integral equation Eq. (4.52) can be solved for the launcher surface [Nei05].

$$j_S(r') = \sum_{n=1}^{N} a_n b_n(r') \tag{4.55}$$

$b_n(r')$ are so-called basis functions and the procedure is called *method of moments (MoM)*. Eq. (4.52) can be rearranged and simplified to:

$$\vec{E}_t^i(r = R_s) = j\frac{\eta}{k_0} \left[k_0^2 \oint_S \sum_{n=1}^{N} a_n b_n(r') g(r, r') ds' \right.$$
$$\left. + \nabla \oint_S \nabla' \cdot \sum_{n=1}^{N} a_n b_n(r') g(R_s, r') ds' \right] \tag{4.56}$$

In a next step, the scattered field $\vec{E}^s$ can be calculated based on the known surface current. The coordinates x, y and z describe again the point of view and x', y' and z' the points on the surface of the conductor which have to be integrated.

$$\vec{E}^s(x, y, z) = -j\frac{1}{4\pi\omega\varepsilon_0} \left[\omega^2 \mu_0 \varepsilon_0 \oint_S \vec{j}_S(x', y', z') \frac{e^{-jk_0 R}}{R} ds' \right.$$
$$\left. + \nabla \oint_S \nabla' \cdot \vec{j}_S(x', y', z') \frac{e^{-jk_0 R}}{R} ds' \right] \tag{4.57}$$

The equation for the scattered wave can again be simplified to:

$$\vec{E}^s(r) = -j\frac{\eta}{k_0}\left[k_0^2 \oint_S \vec{j}_S(r')G(r,r')ds' + \nabla \oint_S \nabla' \cdot \vec{j}_S(r')g(r,r')ds'\right]$$

$$(4.58)$$

The outgoing field from the original optimized surface (Fig. 4.2) is calculated and plotted on a cylinder with a radius of $100\,\text{mm}$ around the launcher. Fig. 4.6(a) shows the $|E_\varphi|^2$-component of both beams with an azimuthal separation of $144°$ and an offset in height. The corresponding phase, as displayed in Fig. 4.6(b), has a constant (linear) aperture at the beams' positions. This phase distribution is characteristic for beams with a high vector correlating coefficient in relation to a Gaussian like aperture.

Fig. 4.7(a) and 4.7(b) show the field and phase distribution of the simplified wall using a two-dimensional filter function according to Sec. 4.4.2. Within these plots, the utilized Chebyshev low-pass filter is set to $\omega_p = 0.3 \cdot \omega_n$ and $\omega_s = 0.4 \cdot \omega_n$ resulting in a filter order of $m = 11$. The pictures can be directly compared to Fig. 4.6(a) and 4.6(b). The filtered surface shows nearly identical amplitude and phase distribution for both output beams in comparison to the original structure.

The full wave three-dimensional analysis of the original and filtered launcher surface verifies the capability to transform the high-order operating mode $TE_{-52,31}$ into two beams with a high vector correlation coefficient with respect to the fundamental Gaussian field distribution. The unchanging high quality of both beams after simplification using two-dimensional filter techniques can be reproduced and verified as well.

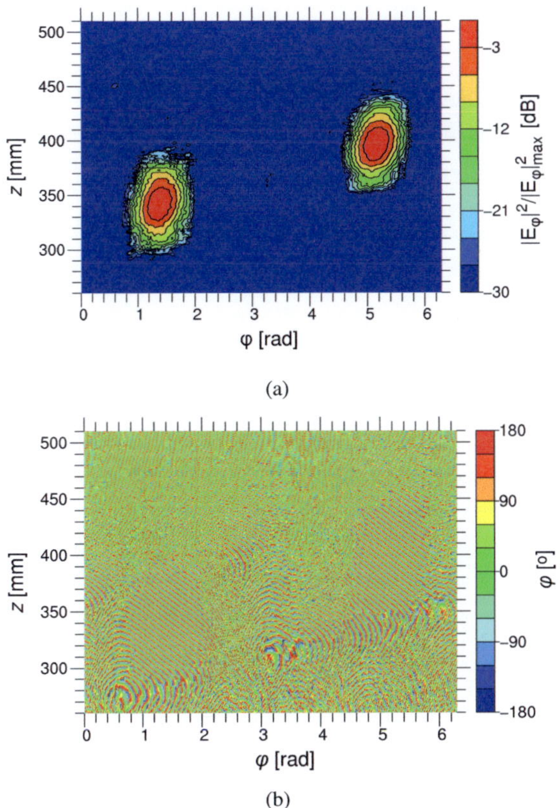

Fig. 4.6.: $|E_\varphi|^2$ (a) and phase (b) of the radiated beams for the launcher surface w/o smoothening

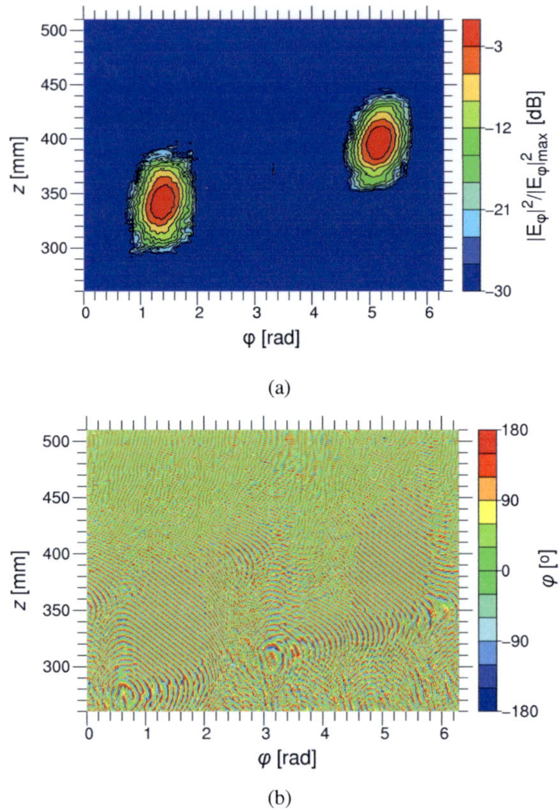

Fig. 4.7.: $|E_\varphi|^2$ (a) and phase (b) of the radiated beams for the surface after two-dimensional low-pass filtering

5. Design studies for the collector

After the electron-cyclotron-interaction in the cavity, the spent electron beam has to be absorbed in the collector. In addition, with a usual interaction efficiency of approximately 35%, the particles after interaction still have a high kinetic energy which needs to be recovered or converted into heat at the collector wall. To guarantee long lifetime of the collector, a limited average wall loading has to be achieved. The collector can be manufactured using pure or dispersion strengthened copper, widely known as *Glidcop*. Glidcop is a registered trademark name, that refers to a family of copper-based metal matrix composite (MMC) alloys mixed primarily with aluminum oxide ceramic particles. The addition of a small amount of aluminum oxide has minuscule effects on the performance of copper (such as small decrease in thermal and electrical conductivity), but greatly increases the material's resistance to thermal softening and enhances strongly the elevated temperature strength. The main material properties for different Glidcop composites can be found in Appx. A.1. The collector needs to be cooled from its outside using techniques with a sufficient heat exchange coefficient to counter the high thermal loading. Usually hyper-vapotron cooling is utilized to achieve high reliability of the collector structure. The water flow is aligned perpendicular to grooves in order to achieve a high fluid turbulence and enable the liquid to vaporize with a high heat exchange coefficient. For long lifetime, the maximum acceptable wall loading for pure copper is $500\,\mathrm{W/cm^2}$ and $1000\,\mathrm{W/cm^2}$ for Glidcop due to the different materials' tensile strengths (see Tab. A.1).

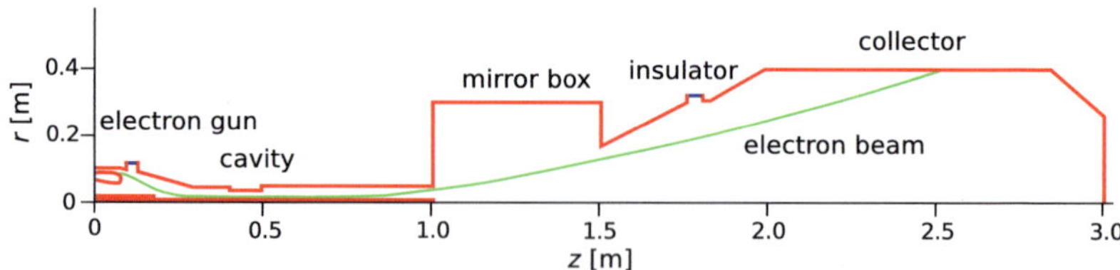

Fig. 5.1.: Complete simulation setup - from triode-type MIG to mirror box and collector

5.1. Possible optimization steps for the collector

Fig. 5.1 shows the complete gyrotron simulation setup for the electron trajectories starting from the triode-type magnetron injection gun to the collector. The mirrorbox is assumed to have a height of $500\,\mathrm{mm}$ in z-direction. The electron trajectories up to the collector input (at $z \approx 1.5\,\mathrm{m}$) only have to be calculated once, because the conditions in this region only change marginally during the collector optimization process. The beam width at the collector wall, without any magnetic sweeping as described later, is about $50\,\mathrm{mm}$ in the vertical dimension, leading to a power density at nominal operation of $> 4\,\mathrm{kW/cm^2}$ which is unacceptably high. The main challenge of the collector optimization is to decrease the average and peak values for the wall loading on the collector surface to acceptable limits.

The classical and well established approach to achieve an advantageous distribution of the wall loading is the utilization of several magnetic sweeping coils around the collector. The electron beam is swept along the collector surface and the size of the absorbing area is rapidly increased. The commonly utilized approaches are combined longitudinal and transversal field sweeping systems as sketched in Fig. 5.2 [DAAea02]. For the longitudinal sweeping, ring shaped magnetic coils around the collector are used to shift the beam along the z-direction. The longitudinal sweeping has a relatively weak electrical efficiency because the copper collector represents a single turn, short-circuited coil, which is shielding the sweep magnetic field very efficiently (typically around 80%). The sweeping is thus restricted to low frequencies in the range of several Hz. As a consequence, powerful AC-power supplies in connection with large, water-cooled sweeping coils are required to provide the necessary capability.

The transversal sweeping system usually consists of three pairs of coils, which are powered with a tree-phase AC-supply [SID$^+$07]. The electron's intersection with the collector wall becomes an ellipse as shown in Fig. 5.2 (right side), which rotates corresponding to the sweeping frequency. To guarantee sufficient penetration of the generated alternating field, the corresponding parts of the tube can be produced using materials with slightly lower electrical conductivity like stainless-steel. For both types of sweeping methods, two main peaks of the wall loading at the upper and lower turning point of the electron beam along the collector surface occur. For the transversal sweeping, the upper maximum can be smoothed by a proper choice of parameters, but the lower one usually remains and no advantage is achieved with respect to the limits, which have to be set to the collector operation. The beam spreading by the transverse magnetic field is more efficient in the upper section of the collector, whereas in the lower area the magnetic field remains stiff and the beam well collimated. Furthermore, time dependent calculations show that the beam resides longer at the lower turning point with a steeper angle of incidence to the collector surface. The power peaking is thus a general feature of both sweeping systems and determines the overall collector capability [ESL$^+$07].

In Fig. 5.3 the power density on the surface is plotted for the non-optimized 4 MW case using six longitudinal sweeping coils. Each coil is assumed to have 60 windings powered by an AC-current of 80 A and 7.0 Hz. The beam is swept along a broad area of approximately 60 cm in z-direction. Two main peaks occur at the lower and higher turning point of the electrons, which greatly exceed the maximal admissible values.

These peaks are generated due to the relative slow sweeping velocity of the beam as a consequence of the sinusoidal form of the currents in the longitudinal coils. Even if a sawtooth like function for the sweeping current is applied, only the low Fourier components can penetrate through the collector wall and the power peaking remains accordingly.

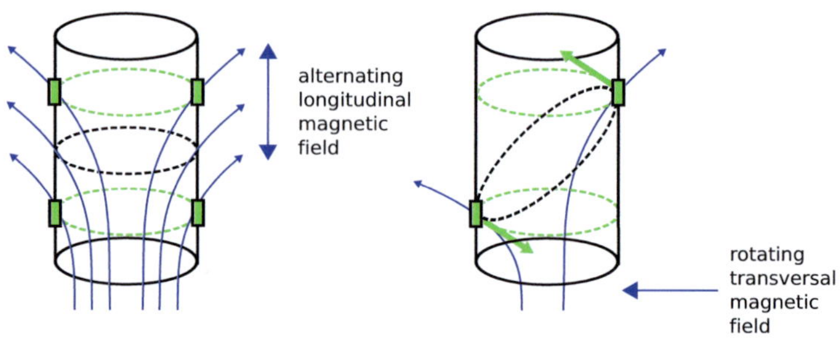

Fig. 5.2.: Principle sketch of longitudinal and transversal collector sweeping systems

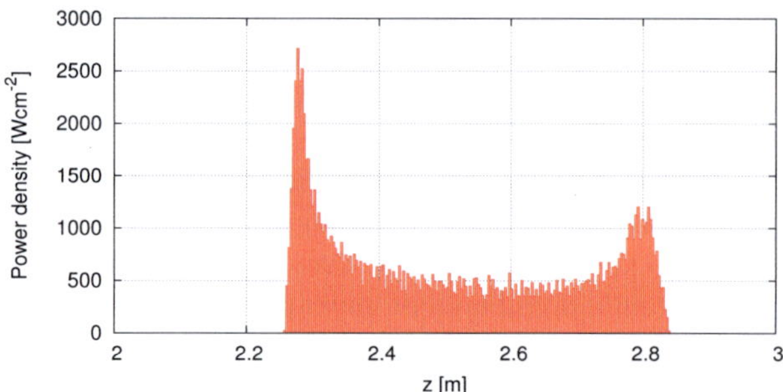

Fig. 5.3.: Loading at non optimized collector surface with six longitudinal sweeping coils (60 windings, 80 A, 7.0 Hz)

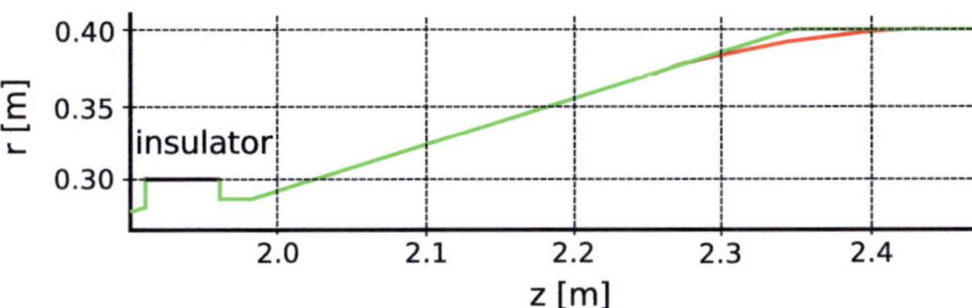

Fig. 5.4.: Lower collector section with smoothed transitions (red) towards the cylindrical upper section; insulator implemented for depression voltage

In addition, the high value at the lower turning point is originated by the steep angle of the electron beam hitting the collector surface. To decrease this peak power density, the shape of the collector surface can be improved around this critical spot. An angled surface with a smoothed section towards the cylindrical collector region increases the relative surface area for absorption. An example for a possible alignment of the surface shape can be seen in Fig. 5.4. The picture shows the lower collector section with the insulator to apply the depression voltages and a straight (green) and optimized (red) surface shape. The angle of the wall relative to the beam needs to be as low as possible and the position and radius of the smoothing towards the cylindrical part have to be chosen accordingly. For a robust design, a possible variation of the alignment of the electron beam due to magnetic stray fields has to be considered.

A further possibility to decrease the peak wall loading at both turning points of the electron beam is the modulation of the magnitude of the current within the longitudinal sweeping coils. This so-called *wobbling* slightly shifts the position of the critical area during every cycle. The amount and velocity of the shift can be easily adapted and optimized. The applied current in every longitudinal sweeping coil $I_{\mathrm{s,n}}$ is split into a DC-$I_{\mathrm{s,n,DC}}$ and an AC-part $I_{\mathrm{s,n,AC}}$. The magnitude of the AC-part is again split into a fixed $I_{\mathrm{s,n,AC,0}}$ and time-variant $I_{\mathrm{s,n,AC,var}}$ portion. ϕ is the phase of the real and imaginary part of the sweeping current and r describes the ratio between the wobbling and the sweeping frequency.

$$I_{s,n} = I_{s,n,DC} + I_{s,n,AC}(\cos(\phi) + j\sin(\phi))$$
$$= I_{s,n,DC} + (I_{s,n,AC,0} + I_{s,n,AC,var} \cdot \sin(r\phi))$$
$$\cdot (\cos(\phi) + j\sin(\phi)) \qquad (5.1)$$

5.2. Determination of the electron trajectories towards the collector

For reasonable collector studies it is necessary to determine the exact velocity distribution of the spent electrons entering the collector. The distribution of the particles and their trajectories are calculated using the designs for the triode-type gun, the coaxial cavity and the launcher as described within this work. The mirrorbox is assumed to have a height along the gyrotron z-axis of $500\,\mathrm{mm}$.

As a first step, the velocities of the electrons at the cavity exit are calculated assuming stationary self-consistent conditions. A standard deviation of 5% of the perpendicular velocity component $\beta_\perp$ at the cavity input is assumed. The simulation of the triode-type MIG from Sec. 3.6 proposed a spread of $\Delta\beta_{\perp,rms} \approx 1.16\%$, as a consequence the choice should be not critical. A more detailed distribution can be achieved, if the part of trajectories between the emitter and the cavity input is also included in the consideration. For simplification and the negligibly low enhancement for the collector design this section is neglected.

The trajectories of the electrons after the cavity through the mirror box and towards collector input is calculated using the PIC-code as described in Sec. 3.4. After this step, the velocity distribution at the input of the collector is known and can be used for the following simulations.

The different distribution of $\beta_\perp$ for 300 macroparticles can be found in Fig. 5.5. The parallel velocity distribution of the particles is assumed to be constant under consideration of the adiabatic approximation.

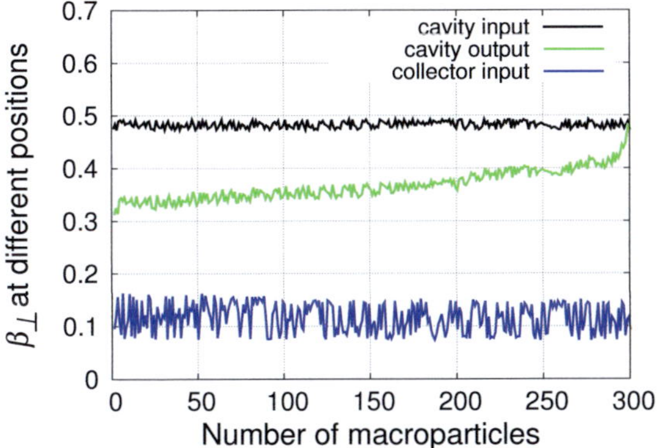

Fig. 5.5.: $\beta_\perp$ at different positions along the beam path for an arbitrary number of 300 macroparticles (random sequence)

5.3. Design of a copper collector

For reliable operation and long collector lifetime, the wall loading using copper as absorption material should be $< 500\,\mathrm{W/cm^2}$. As described above, a transversal sweeping system does not significantly decrease the peak loading at the lower turning point of the beam. In the shown design process, a transverse sweeping system is therefore not considered. The collector layouts are optimized mainly by introducing efficient longitudinal sweeping and a proper surface shape.

The initial collector radius is chosen to be 400 mm. Ten sweeping coils for longitudinal sweeping are aligned along the collector's z-axis starting at a relatively low position to get sensitive influence around the turning point of the beam. Each coil is assumed to have a length of 150 mm in axial direction, a radial width of 40 mm and 60 windings. Several sweeping current distributions are investigated. With a higher constant part of the current amplitude $I_{\mathrm{s,n,AC,0}}$ the sweeping area can be broadened. This is sufficient to reduce the average wall loading between the turning points, which is normally not problematic. As a disadvantage, the wall loading at both turning points is strongly increased and high thermal stresses are introduced. It is not possible to decrease the peak wall loading by utilization of higher variable (wobbling) portions of the sweeping current $I_{s,n,AC,var}$. Higher $I_{s,n,AC,var}$ increases the modulated area at both ends of the sweeping region, but does not reduce the peak wall loading. Consequently, a trade-off between peak and average value is necessary.

To reach the desired objectives, the maximal radius of the collector needs to be set to a relatively large value of 400 mm. The wall loading without depression voltage of the optimized collector setup using copper as surface material is shown in Fig. 5.6. The design considerations include three generations of secondary electrons emitted from the collector wall. The total absorbed power of the spent electrons without depression voltage was calculated to 9.57 MW. The loading has a relatively low peak at 2.28 m with approximately 1050 W/cm^2. The current of the collector sweeping coils is wobbled with $r = 1/7$. With higher constant part of the current amplitude $I_{s,n,AC,0}$ the sweeping area can be broadened, but the values at the turning points remain problematic. A higher variable portion $I_{s,n,AC,var}$ increases the modulated area at both ends of the sweeping region, but does not reduce the peak wall loading as well. These effects are in agreement with the expected behavior as described above and in Sec. 5.1.

The total absorbed power inside the collector decreases with the applied depression voltage. A distribution of the wall loading which is within the acceptable limits is only achievable with high depression. The total absorbed power inside the collector decreases to 4.59 MW for $V_{depr} = 50$ kV. This value is also close to the upper limit for reflected electrons to occur with the considerations for the electron velocity distribution as mentioned in Sec. 5.2. Fig. 5.7 shows the loading on the surface after optimization with improved smooth transition for $V_{depr} = 50$ kV. The main peaks are limited to 400 W/cm^2 and 250 W/cm^2 respectively. In conclusion, a high depression voltage is necessary to achieve the desired peak wall loading in order to guarantee long and reliable collector life-time. This demands a high quality electron beam with low spread of the particles' velocity components. In addition, the distribution of the wall loading is sensitive to deviations in the magnetic sweeping field, which makes the designs sensitive to magnetic stray fields.

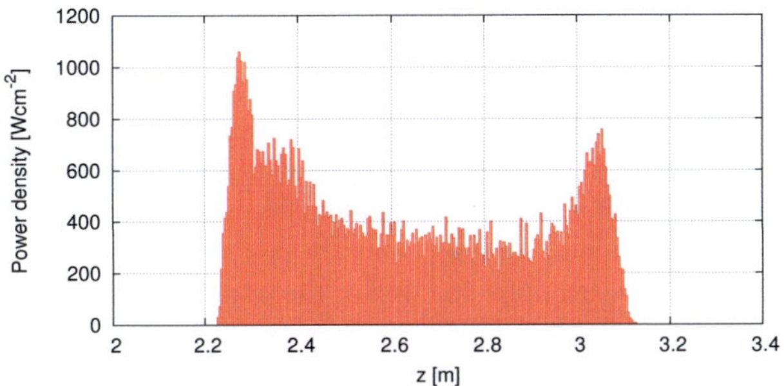

Fig. 5.6.: Wall loading on copper collector after optimization with $V_{\mathrm{depr}} = 0\,\mathrm{kV}$ (9.57 MW of total absorbed power)

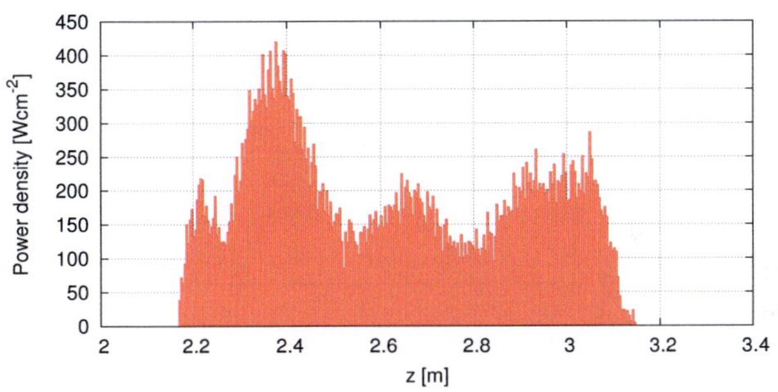

Fig. 5.7.: Wall loading on copper collector after optimization with $V_{\mathrm{depr}} = 50\,\mathrm{kV}$ (4.59 MW of total absorbed power)

5.4. Design of a Glidcop collector

As explained in Sec. 5.3, it is possible to find a collector design using copper as absorbing material with a maximum wall loading of $< 500\,\mathrm{W/cm^2}$. The copper collector has an inner radius of $400\,\mathrm{mm}$ and the electrons are absorbed within a range of approximately $1\,\mathrm{m}$ along the z-axis. This relatively massive collector design would lead to high requirements for the solidness and robustness of the mechanical layout of the complete tube. In addition, the desired wall loading can only be reached with a high depression voltage to recover a large fraction of the remaining energy in the spent electrons.

For dispersion strengthened copper (Glidcop) the upper limit for the wall loading is approximately $1000\,\mathrm{W/cm^2}$, which allows a more compact collector design with respect to radius and axial length. For an initial setup, the radius of a Glidcop collector is chosen to be $300\,\mathrm{mm}$. For the following considerations a model with eight sweeping coils, each with 60 windings, is utilized. The coils are aligned at relatively low z-values around the collector input to get sensitive influence on the lower turning point of the swept beam. The shape of the collector surface and the wobbled sweeping currents are optimized. It is possible to find an optimum for the wall loading of the wobbled coil currents as shown in Tab. 5.1. The sweeping frequency is $7.0\,\mathrm{Hz}$ and the amplitude is modulated with $1\,\mathrm{Hz}$ (ratio $r = 1/7$).

	coil #1	#2	#3	#4	#5	#6	#7	#8
$I_{\mathrm{s,n,AC,0}}$ [A]	60	60	100	100	120	120	120	120
$I_{\mathrm{s,n,AC,var}}$ [A]	20	20	20	20	20	20	20	20

Tab. 5.1.: Optimized wobbled sweeping AC-currents with $r = 1/7$ and $7.0\,\mathrm{Hz}$

For this setup, the wall loading without depression and with $V_{\mathrm{depr}} = 50.0\,\mathrm{kV}$ is shown in Fig. 5.8 and 5.9. The necessary axial length of the absorbing surface is $< 800\,\mathrm{mm}$. The loading is smoothly distributed along the collector surface and no steep gradient is generated. As a consequence, the thermal stresses during operation are significantly lower in comparison to the copper design from Sec. 5.3 and the requirements to the material strength are reduced. Already with $V_{\mathrm{depr}} = 30\,\mathrm{kV}$ the limit of $< 1000\,\mathrm{W/cm^2}$ can be achieved. The design has therefore lower requirements towards the required quality of the electron beam.

For a robust collector design it is necessary to guarantee a relatively high separation of the beam towards the collector wall at positions even far from the area of absorption for every phase position during one sweeping cycle. Magnetic stray fields at the place of the gyrotron's installation might divert the electrons trajectories in a way that some particles hit the collector wall outside the desired area. To overcome this issue, the optimized design has a high minimal distance of the electron beam of approximately 24 mm towards the wall along its path during one cycle.

Analog to Sec. 5.3, the wall loading is determined considering three generations of secondary electrons. For the characterization of secondary particles in Glidcop the same emission models as for copper are utilized. Due to the similarity of the materials, the introduced inaccuracy is assumed to be small. The peaks at both turning points are wiped due to the introduction of secondary emitted particles and the values are lowered by approximately $150\,\mathrm{W/cm^2}$.

To estimate the required capabilities of the necessary power supply for the sweeping coils, their reactive and apparent power is calculated as displayed in Tab. 5.2. An apparent power up to 9 kVAs represents no special requirement towards the power supply for the longitudinal collector sweeping.

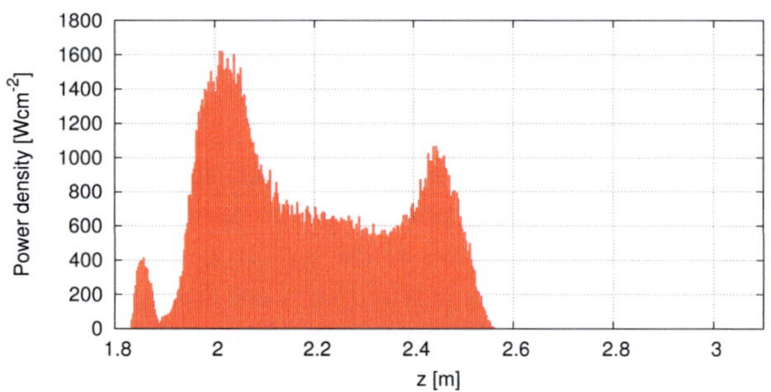

Fig. 5.8.: Wall loading with optimized sweeping currents and $V_{\mathrm{depr}} = 0\,\mathrm{kV}$ (9.57 MW of total absorbed power)

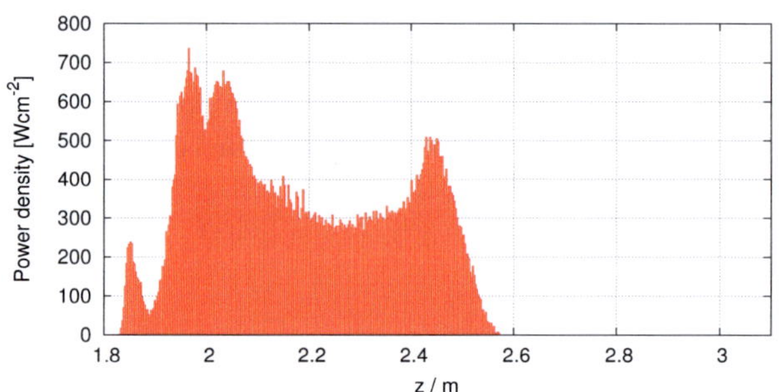

Fig. 5.9.: Wall loading with optimized sweeping currents and $V_{\mathrm{depr}} = 50\,\mathrm{kV}$ (4.59 MW of total absorbed power)

coil #	$I_{\text{AC},0}$ [A]	P [W]	Q [VAr]	S [VA]
coil 1	60	155.3	438.7	465.3
coil 2	60	212.8	471.9	517.6
coil 3	100	647.7	930.4	1133.7
coil 4	100	685.3	937.7	1161.4
coil 5	120	691.8	939.4	1166.7
coil 6	120	692.7	942.7	1169.8
coil 7	120	690.9	982.1	1200.8
coil 8	120	562.1	1152.3	1282.1
sum	—	4338.6	6795.2	8097.4

Tab. 5.2.: Real, reactive and apparent power consumption of eight sweeping coils for the Glidcop collector design (numbering from lowest to highest coil in z-direction)

6. Thermo-mechanical studies of long-pulse operation effects

For long-pulse operation of a mega-watt gyrotron, thermo-mechanical challenges become key issues to guarantee stable and reliable operation conditions. In order to assure high performance of the several components after heat treatment, the cooling techniques for the components have to be designed carefully. In the following chapter, several thermo-mechanical approaches are studied in order to identify and characterize critical issues within the component designs. During gyrotron startup, the cavity is deformed due to thermal treatment which leads to a shift of the output frequency and the quality factor. An estimation of the expected surface temperature of the cavity can offer an insight whether the applied cooling techniques are sufficient and the oncoming enlargement of the resonator is critical. The impedance corrugation of the coaxial insert, which is used for improved mode competition, consist of fine grooves. The structure is deformed during operation and its desired influence might be reduced. A sensitive wall structure can also be found inside the quasi-optical launcher to convert the high-order volume mode into two Gaussian beams. The strongly focused beams lead to a high wall loading on the surface towards the output of the launcher. The converted beams are defocused during long-pulse operation and thus the overall efficiency of the quasi-optical system is reduced and the amount of stray radiation is increased.

All thermo-mechanical calculations are accomplished employing the commercially available finite-element software ANSYS in Version 12.1.

6.1. Deformation of the coaxial cavity and frequency shift

During the startup of a gyrotron pulse, the cavity is deformed due to thermal expansion until it reaches its thermal steady state. The field inside the cavity and the corresponding wall loading heat the structure from the inside, which leads to a shift of the output frequency towards lower values. The cavity enlargement also goes along with a change of the quality factor, which might reduce the stability or efficiency of the interaction. On the outer sections of the cavity several different cooling techniques are applied depending on the necessary heat exchange coefficient. In the following sections, the basic considerations and simplifications are introduced and the steady state temperature distribution and corresponding cavity expansion are determined. If the cavity's enlargement is known, the design of the coaxial cavity can be adjusted in order to compensate the expected frequency shift and quality factor detuning.

6.1.1. Considerations and theory for the cavity wall loading

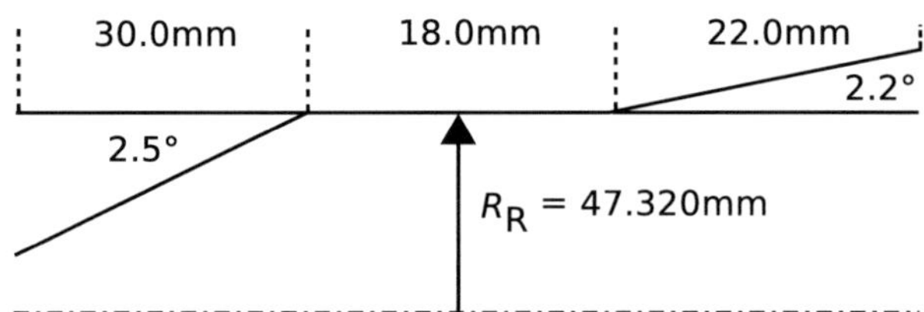

Fig. 6.1.: Simplified cavity geometry for thermo-mechanical studies

The main geometrical parameter which determines the output frequency is the radius of the cylindrical resonator section where the longitudinal RF-field profile reaches its maximum. The nonlinear tapers at both ends of the cavity can be replaced by linear ones to simplify the model. The reduced geometry of the interaction cavity, as it is utilized in the thermo-mechanical calculations, is shown in Fig. 6.1. The field profile and the corresponding wall loading of the cavity's inner surface is calculated in the case of ideal copper at room temperature (solid black line $\rho_{R,ideal}$ in Fig. 6.2) using the self-consistent stationary approach as described in Sec. 2.1. The peak wall loading is $1.0\,\mathrm{kW/cm^2}$ and the data is related to a generated RF-power of $4.5\,\mathrm{MW}$.

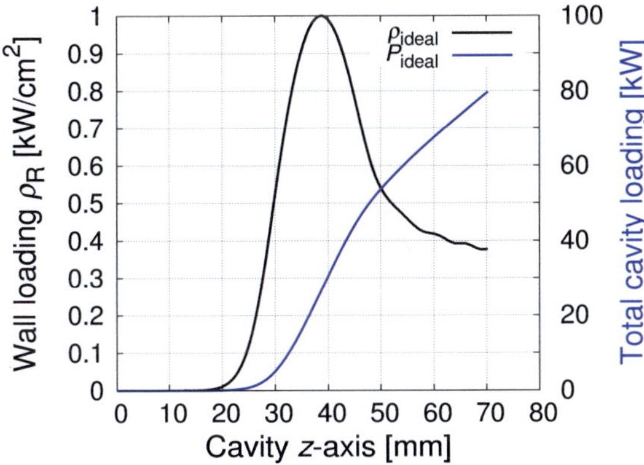

Fig. 6.2.: Power density and total wall loading for ideal copper along cavity z-axis

The power density has to be scaled in order to take into account the conductivity of Glidcop, instead of ideal oxygen-free high thermal conductivity copper (OFHC), and the surface roughness due to the machining process. Glidcop is up to now the best compromise between mechanical properties and thermal behavior. The conductivity of Glidcop as a function of the temperature can be incorporated in the power density profile with the following empirical equation [Clo01]:

$$\frac{\sigma_{\mathrm{Glidcop}}}{\sigma_{\mathrm{Cu,ideal,273K}}} = 7\cdot 10^{-7}T^2 - 0.0011T + 0.9447 \quad \text{with} \quad \text{T in } {}^\circ\text{C} \quad (6.1)$$

The corresponding plot can be found in Fig. 6.3(a). The surface roughness is represented by a constant coefficient k_{sr}, which is usually chosen to be $k_{sr} = 1.5$ for real dispersion strengthened copper. Consequently the total wall loading for machined Glidcop and ideal copper are related by:

$$P_{\text{Glidcop}} = P_{\text{Cu}} \cdot k_{sr} \sqrt{\frac{\sigma_{\text{Cu}}}{\sigma_{\text{Glidcop}}}} \qquad (6.2)$$

The peak power density of the wall loading is thus theoretically increased by approximately 70%. Within real measured approaches, the dissipated peak power is nearly doubled in comparison to ideal assumptions [Gei91]. In this work, $1.7\,\text{kW/cm}^2$, which is close to the theoretical value from Eq. (6.2), is assumed to be dissipated in the middle of the cavity. This is slightly less than the value considered before in Tab. 2.4.
Fig. 6.3(b) shows the total employed power for the coaxial 4 MW cavity dependent on the surface roughness k_{sr} of Glidcop. For a given roughness $k_{sr} = 1.5$ a power dissipation of 136.2 kW inside the cavity is achieved.

6.1.2. Steady state temperature distribution

Steady state thermal calculations are performed with the previously derived data from Sec. 6.1.1 on the simplified geometry shown in Fig. 6.1. The radial thickness of the material (dispersion strengthened copper) is constant at 2.0 mm for all cavity sections. The simulated geometry has a total length of 70 mm in z-direction. *Raschig rings* are utilized to achieve highly efficient cooling of the cavity. Raschig rings are pieces of copper tubes (approximately equal in length and diameter) used in large numbers to strongly increase the turbulence within the cooling fluid and the corresponding surface area. The rings are brazed to the outside of the cylindrical part of the cavity and the uptaper section providing a porous cooling structure. The radial width of the cooling structure is assumed to be 30.0 mm to guarantee high flow rates of the liquid. The boundary conditions for the simulations can be found in Tab. 6.1 and the heat exchange coefficients h_c are gathered from [BL07].

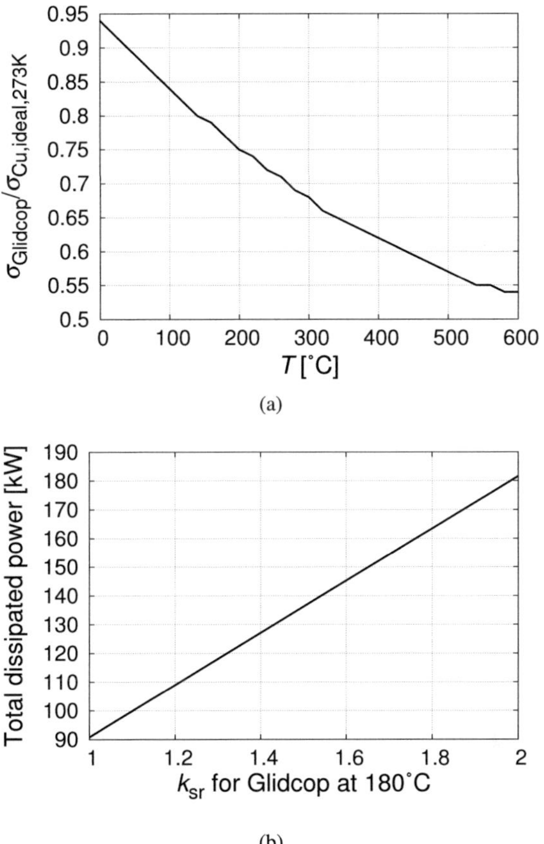

(a)

(b)

Fig. 6.3.: σ_{Glidcop} versus temperature T (a) and total wall loading (b) inside cavity versus k_{sr} at $180\,°C$

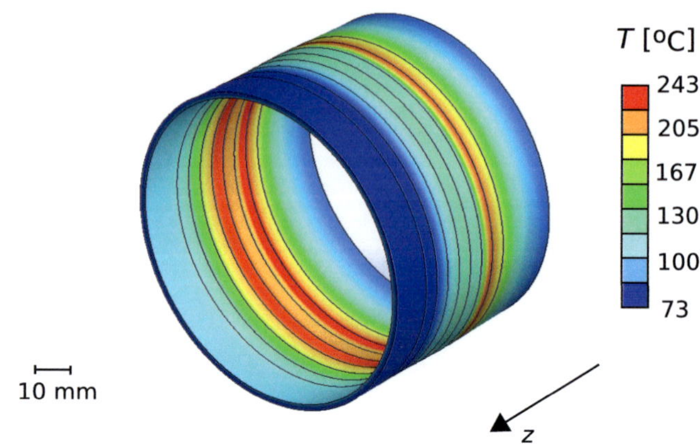

Fig. 6.4.: Steady state temperature distribution at the cavity for Glidcop with $k_{sr} = 1.5$ and $h_c = 0.13\,\mathrm{W}/^\circ\mathrm{Cmm}^2$

Fig. 6.4 shows the hot steady state temperature distribution along the undeformed cavity for a heat exchange coefficient of $h_c = 0.13\,\mathrm{W}/^\circ\mathrm{Cmm}^2$ within the Raschig rings and $k_{sr} = 1.5$ for Glidcop. The inner surface is heated to a maximum temperature of approximately $243\,^\circ\mathrm{C}$. The calculation shows two areas of highest temperature: One at the field maximum within the cylindrical cavity section and the second at the downtaper where the conventional water cooling without Raschig rings is applied. This behavior with two hot areas is not critical and can be avoided by a shift of the Raschig ring structure and its high heat exchange towards the downtaper section. At the end of the cavity uptaper, the temperature decreases rapidly to approximately $73\,^\circ\mathrm{C}$. Fig. 6.5 shows the maximal temperature at the cavity's inner surface depending on the heat exchange coefficient h_c of the Raschig structure. The heat exchange rate for the water cooled parts is kept constant according to Tab. 6.1. The shown maximal temperature depends strongly on the coefficient k_{sr} expressing the surface roughness. The smoothness of the cavity surface has to be checked carefully in order to avoid excessive heat. An even more intense increase of the surface temperature is calculated for lower efficiency h_c of the Raschig structure. As a consequence, the velocity of the cooling water and the alignment of the Raschig rings has to be checked carefully.

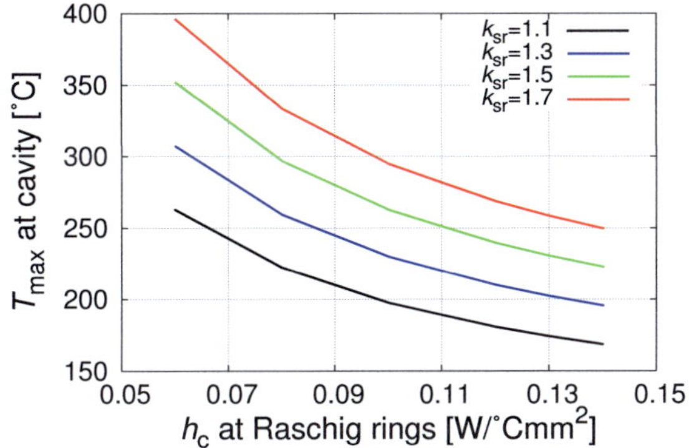

Fig. 6.5.: T_{max} at cavity inner surface versus heat exchange h_c of Raschig structure (for other parameters see Tab. 6.1)

position	boundary condition	h_c [W/$^\circ$Cmm^2]	comment
inside complete	heat flux	–	as described above
outside downtaper	linear convection	0.003	water
outside cavity and uptaper	linear convection	0.13	water & Raschigs

Tab. 6.1.: Boundary conditions for thermal steady state analysis of the cavity

6.1.3. Verification and comparison with experimental data

To verify the determined frequency shift due to the cavity's thermal expansion experimental data from a well known series gyrotron developed for the Wendelstein 7-X stellarator (W7-X) is utilized [Erc02][DAAea02]. Fig. 6.6 shows the frequency shift of series tube #1 and the achievable acceleration voltage V_{acc} over the magnitude of the magnetic field B_{R} in the cavity. The shown data corresponds to an output power of $900\,\mathrm{kW}$ at $140\,\mathrm{GHz}$. The measured frequency shift Δf during long pulse operation is approximately $-250\,\mathrm{MHz}$ over a broad range of operating parameters.

After calculation of the cavity deformation within the finite-element approach, the frequency shift is determined. Fig. 6.7 shows the normalized longitudinal field amplitude along the cavity for different wall loadings ρ_{R} at stationary self-consistent conditions. The deformation of the cavity leads to a diminution of the longitudinal field profile. The design of the W7-X cavity is optimized for $\rho_{\mathrm{R}}{=}1.3\,\mathrm{kW/cm^2}$ at the nominal working point and realistic material properties (rough Glidcop).

As shown in Fig. 6.8, the deformation of the cavity and the frequency shift Δf of the cavity are strongly dependent on the roughness of the cavity's inner surface. For $\rho_{\mathrm{R}} = 1.3\,\mathrm{kW/cm^2}$ and $k_{\mathrm{sr}} = 1.5$ the shift Δf is calculated to be $-213\,\mathrm{MHz}$ for an undeformed cavity output frequency of $140.203\,\mathrm{GHz}$. The results are in good agreement with the data gathered from the experiments and the simplifications for the numerical approach seem to be reasonable. The slightly smaller calculated frequency shift of $-213\,\mathrm{MHz}$ in comparison to the measured value of $-250\,\mathrm{MHz}$ could be an evidence for a higher real surface roughness ($k_{\mathrm{sr}} > 1.5$) or less effective cooling at the Raschig rings structure during gyrotron operation. In addition, the deformed cavity has a higher Q-value as shown in the following section.

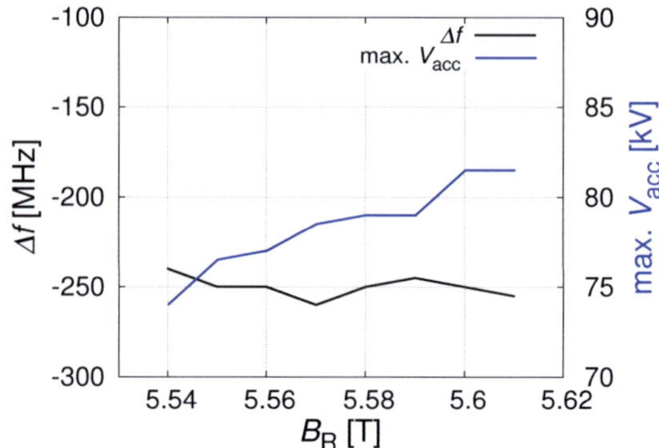

Fig. 6.6.: Δf at startup (black) for W7-X series tube #1 at approximately 900 kW

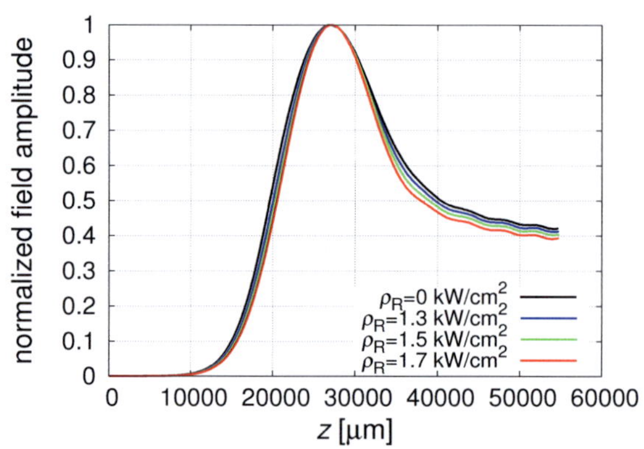

Fig. 6.7.: Normalized field profiles for the W7-X 140 GHz cavity at different wall loadings ρ_R

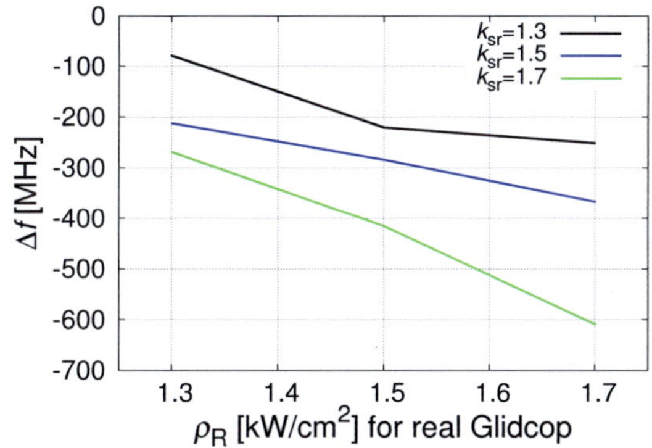

Fig. 6.8.: Calculated frequency shift Δf versus ρ_R for different cavity surface roughnesses

6.1.4. Deformation and frequency shift for the 4 MW 170 GHz cavity

In a next step, the deformation of the optimized cavity from Sec. 2.5 and corresponding frequency change is determined. Fig. 6.9 displays the enlarged cavity for the 4 MW 170 GHz gyrotron at a maximum temperature of 243 °C. As shown in Sec. 6.1.2, the temperature corresponds to a high wall loading of $\rho_R = 1.7\,\mathrm{kW/cm^2}$ at the nominal working point of the possible 4 MW gyrotron. The utilized geometrical parameters for the finite-element setup are given in Tab. 2.3, and the applied heat exchange coefficients again in Tab. 6.1. Like mentioned before, the cooling structure at the downtaper section is assumed to be normal water and Raschig rings are aligned at the cylindrical cavity section and the uptaper.

The maximum deformation on the inside of the cavity in radial direction is approximately $\Delta R_R = 91\,\mu\mathrm{m}$ at the nominal working point. The maximum enlargement is located close to the maximum field amplitude inside the cavity. The deformation at the uptaper section is calculated to be very low and a corresponding change in the performance of the mode conversion properties towards the launcher is not expected.

In Fig. 6.10(a) the maximal temperature T_{max} on the inside of the cavity surface is plotted depending on the heat exchange coefficient of the Raschig structure and the Glidcop surface roughness k_{sr}. In [Clo01] the usual value for h_{c} of the Raschig structure is determined to $0.13\,\mathrm{W/°Cmm^2}$. Fig. 6.10(a) describes a strong dependency of the temperature at the cavity inner surface on the efficiency of the cooling structure as well as the surface roughness. For a non optimal smoothing of the surface its temperature can increase up to $100\,°\mathrm{C}$ in comparison to an ideally plain wall. A weakly aligned Raschig structure in combination with a rough surface could lead to critical conditions which have to be avoided.

The frequency shift after deformation is again calculated using the stationary self-consistent field profile. The corresponding data is plotted in Fig. 6.10(b). A cold and undeformed cavity delivers an output frequency of 170.149 GHz corresponding to a cavity radius of $R_R = 47.320\,\mathrm{mm}$.

The expected frequency shift Δf of the deformed 4 MW 170 GHz cavity for a wall loading of $\rho_R = 1.7\,\mathrm{kW/cm^2}$ and $k_{\mathrm{sr}} = 1.5$ will be approx-

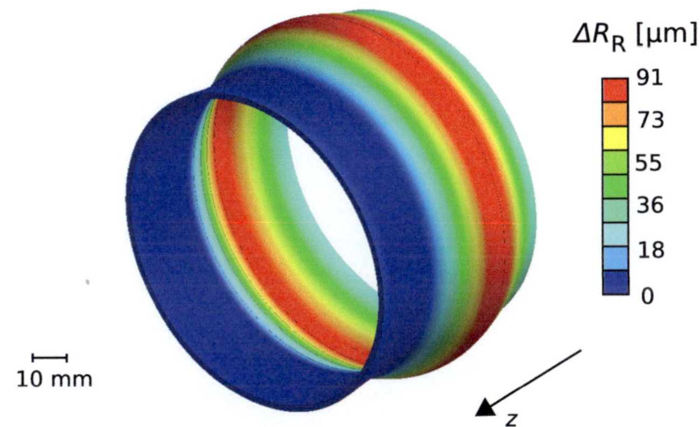

Fig. 6.9.: Radial deformation of the cavity at thermal steady state (strongly magnified) with $T_{\mathrm{max}} = 243\,^{\circ}\mathrm{C}$ on the inner surface

imately $-129\,\mathrm{MHz}$. As described in Sec. 2.5, the slightly lower chosen value for the radius of the optimized coaxial cavity will generate an output frequency of almost exact $170.0\,\mathrm{GHz}$.

As a further consequence, the deformation of the cavity leads to a shift of the resonator's quality factor Q. The amount of energy stored in the resonator as well as the released energy per cycle changes due to the expansion. Fig. 6.11(a) shows Q in self-consistent calculation of the cavity in thermal steady state. The resonator shows an increasing quality factor for reduced efficiency of the cooling structure. The values are again strongly dependent on the roughness of the surface. Already small deviations of the heat exchange coefficient at the cylindrical cavity section lead to tremendous changes in the resonator characteristics. In order to guarantee stable operation, the efficiency of the cavity cooling and the surface roughness have to be checked carefully.

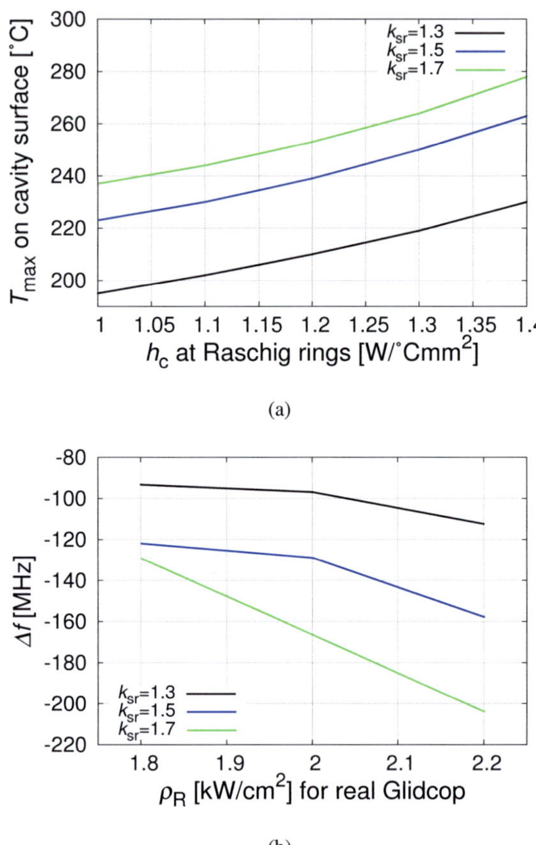

Fig. 6.10.: T_{max} versus h_{c} (a) and corresponding frequency shift Δf versus cavity loading ρ_{R} (b)

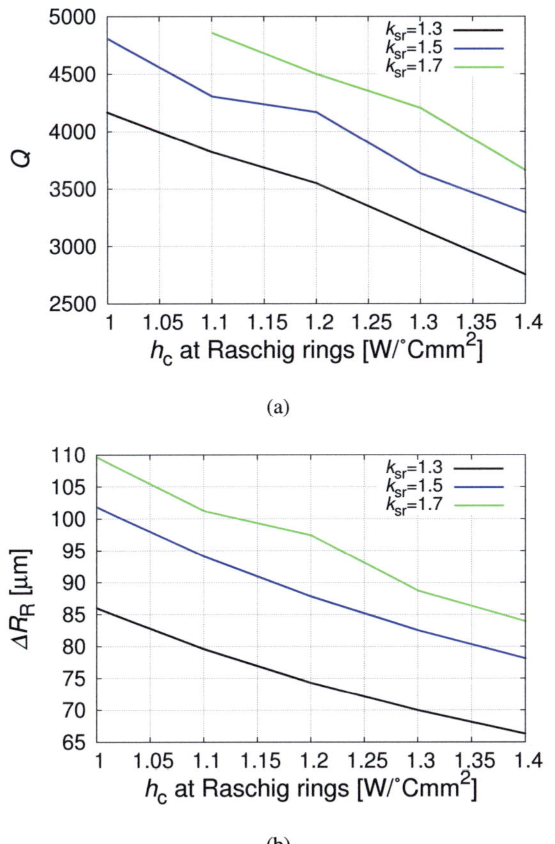

Fig. 6.11.: Quality factor Q (a) and corresponding maximal radial deformation ΔR_R (b) of the cavity in thermal steady state

6.2. Deformation of the impedance corrugation at the coaxial insert

Besides the expansion of the cavity profile, the coaxial insert is heated during long pulse operation. The corresponding deformations are especially critical for the fine grooves of the impedance corrugation on the outer side of the insert. The corrugation is utilized to influence the eigenvalue curves of the modes to improve mode competition. In a coaxial geometry, the eigenvalue $\chi_{m,p}$ of a $TE_{m,p}$ mode depends on the ratio of the outer to the inner conductor radius $C = R_R/R_i$. With an appropriately tapered outer wall and coaxial insert, $\chi_{m,p}(z)$ is not constant, and a selective influence on the diffractive quality factor Q_{diff} of different modes is possible. Details on the mechanism introduced by impedance corrugation in order to improve the mode competition are gathered in Sec. 1.3.3 based on [Ker96]. The coaxial insert for a 4 MW 170 GHz gyrotron, as shown in Sec. 2.5, has an outer radius of $R_i = 13.0$ mm at the cavity center. In order to avoid mode conversion, the number of grooves has to exceed approximately two times the azimuthal index m of the operating mode. Thus, for the $TE_{-52,31}$ mode, an impedance corrugation has to consist at least out of 104 grooves with a width of 0.39 mm or smaller. For the corresponding fine structure, small temperature changes could lead to distortions which reduce the efficiency of the corrugation significantly. Within the next sections, the different theories to describe the wall loading on the coaxial insert are introduced and possible deformations during long pulse operation are discussed.

6.2.1. SIM and SIE theory

To characterize the loading on the corrugated coaxial insert, several theoretical descriptions exist [DZ04]. Two of these approaches, which are utilized in this work, are namely the surface impedance model (SIM) and the single integral equation (SIE) model.
Within the surface impedance model, the field inside the corrugation's grooves is assumed to be homogeneous, although it can vary from one groove to another according to the mode's azimuthal field distribution. This

150

approximation is based on the assumption that the width of the corrugations is smaller than half of the oscillating wavelength ($l < \lambda/2$). With the following equation the eigenvalue $\chi_{m,p}$ of the $TE_{m,p}$ mode can be determined:

$$J'_m(\chi_k)\left[N'_m\left(\frac{\chi_k}{C}\right) + W N_m\left(\frac{\chi_k}{C}\right)\right]$$
$$- N'_m(\chi_k)\left[J'_m\left(\frac{\chi_k}{C}\right) + W J_m\left(\frac{\chi_k}{C}\right)\right] = 0 \qquad (6.3)$$

The index k relates the single values to one specific mode (mode index). J_m and N_m are again the Bessel and Neumann functions and W is the corrugation parameter. W is usually defined as:

$$W = \frac{l}{s}\tan\left(\frac{2\pi}{\lambda}d\right) \qquad (6.4)$$

l, s and d are the geometrical parameters to describe the impedance corrugation as shown in Fig. 1.9. For every specific value $C(z)$, Eq. (6.3) can be solved for $\chi_{m,p}$ numerically along the cavity profile.

Within SIM the densities of ohmic losses on the different walls (top, bottom, side) are given in the following formulas [Ker96]:

$$\rho_{\text{SIM,top}} = \frac{\delta C_k^2}{4\omega\mu_0}\cdot\left[J_m\left(\frac{\chi_k}{C}\right) - \frac{J'_m(\chi_k)}{N'_m(\chi_k)}N_m\left(\frac{\chi_k}{C}\right)\right]^2$$
$$\cdot\left[\frac{m^2 C^2}{R_R^2}\left|\frac{\partial}{\partial z}f_k(z,t)\right|^2 + \frac{\chi_k^4}{R_R^4}|f_k(z,t)|^2\right] \qquad (6.5)$$

$$\rho_{\text{SIM,bottom}} = \frac{\delta}{4\omega\mu_0}\frac{\chi_k^4}{R_R^4}\left|A^s_{0,n}\right|^2|f_k(z,t)|^2 \qquad (6.6)$$

$$\rho_{\text{SIM,side}} = \frac{\delta}{4\omega\mu_0}\frac{\chi_k^4}{R_R^2}\left|A^s_{0,n}\right|^2|f_k(z,t)|^2\cos^2(k_{\perp,k}(r - R_i + d)) \qquad (6.7)$$

δ denotes the penetration depth and C_k is a constant for normalization.

$$C_k = \left\{ \cfrac{\cfrac{\pi^{-1}}{(\chi_k^2 - m^2)\, Z_m^2(\chi_k)} \quad \cdots}{-\left[\frac{\chi_k^2}{C^2}(1 + W^2) - m^2 - \frac{\chi_k^2}{C}\left(\frac{W}{\chi_k} + \left(\frac{l}{s} + \frac{s}{l}w^2\right)\frac{d}{R}\right)\right] Z_m^2\left(\frac{\chi_k}{C}\right)} \right\}^{1/2} \tag{6.8}$$

with the linear combination Z_m of Bessel and Neumann functions.

$$Z_m(\chi) = J_m(\chi) - \frac{J_m'(\chi_k)}{N_k'(\chi_k)} N_m(\chi) \tag{6.9}$$

The amplitude of the normalized field component inside the grooves $A_{0,n}^S$ can be determined due to the continuity of the H_z-component at $r = R_i$.

$$\begin{aligned}
H_z = H_z^s &= -j\frac{k_{\perp,k}^2}{\omega\mu_0}\left[J_m\left(\frac{\chi_k}{C}\right) - \frac{J_m'(\chi_k)}{N_m'(\chi_k)} N_m\left(\frac{\chi_k}{C}\right)\right] C_k e^{jm\varphi_s} \\
&= j\frac{k_{\perp,k}^2}{\omega\mu_0} A_{0,n}^S \cos(k_{\perp,k}d)
\end{aligned} \tag{6.10}$$

with

$$A_{0,n}^s = \frac{-\left[J_m\left(\frac{\chi_k}{C}\right) - \frac{J_m'(\chi_k)}{N_m'(\chi_k)} N_m\left(\frac{\chi_k}{C}\right)\right]}{\cos(k_{\perp,k}d)} C_k e^{jm\varphi_s} \tag{6.11}$$

φ_s is the azimuthal solid angle which slips over the length s according to Fig. 1.9. In consequence, the loss density increases from a fixed value on the upper side of the corrugation with a $\cos^2$-characteristic towards the ground of the grooves. With a properly chosen corrugation depth ($d = \lambda/4$), the loading on the top side of the bulks vanishes. Within SIM higher spatial harmonics are neglected, which may lead to uncertainties in evaluating the local density of ohmic losses as well as their average values given in Eq. (6.5) to (6.7).

A more comprehensive method is based on a singular integral equation, which is derived for the total field without any simplifying assumptions. A detailed mathematical deviation of SIE is given in [DZ04].

The losses can be expressed using a membrane function $u(r, \varphi)$ by superposition of spatial harmonics.

$$u(r, \varphi) = \sum_{n=-\infty}^{\infty} A_n f_n(r) e^{jk_n \varphi} \tag{6.12}$$

A is an overall normalization factor and $f(z)$ is again the longitudinal component of the RF-field. The local density of losses $\rho(z)$ are determined by the output power and the diffractive quality factor.

$$\rho(z) = \frac{1}{2} \delta k_0^2 \frac{|f(z)^2|}{\int_0^{z_{out}} |f(z)|^2 dz} |u(r, \varphi)|^2 Q_{\mathrm{dif}} P_{\mathrm{out}} \tag{6.13}$$

The corresponding loadings on the different walls are given in the following equations:

$$\rho_{\mathrm{SIE,top}}(z) = \frac{\rho(z)}{|u(r, \varphi)|^2} \cdot \frac{2}{\varphi_s - \varphi_l} \int_{\varphi_l/2}^{\varphi_s/2} |u(r, \varphi)|^2|_{r=R_i} \, d\varphi \tag{6.14}$$

$$\rho_{\mathrm{SIE,bottom}}(z) = \frac{\rho(z)}{|u(r, \varphi)|^2} \cdot \frac{1}{d} \int_{R_i-d}^{R_i} |u(r, \varphi)|^2|_{\varphi=\varphi_l/2} \, dr \tag{6.15}$$

$$\rho_{\mathrm{SIE,side}}(R, z) = \frac{\rho(z)}{|u(r, \varphi)|^2} \cdot \frac{1}{\varphi_l} \int_{-\varphi_l/2}^{\varphi_l/2} |u(r, \varphi)|^2|_{r=R_i-d} \, d\varphi \tag{6.16}$$

Analog to the description above, φ_s and φ_l are the azimuthal solid angles defined from the center of the coaxial insert to the characteristic corrugation lengths s and l. Depending on the geometrical corrugation parameters, the higher spatial harmonics in Eq. (6.12) can either increase or decrease the ohmic losses. The ratios of the wall loading given on the several walls normalized to the bottom surface of the corrugation are given in Tab. 6.2. These ratios are utilized in the following thermo-mechanical description of the insert's deformation.

Generally, SIE predicts lower average wall loading in comparison to SIM for an equal field amplitude. In addition, the loading on the top side does not vanish for an appropriate choice of corrugation parameters within SIE. In the following calculations of the deformations on the impedance corrugation, both models are utilized and the results are compared respectively.

wall	SIM	SIE
ρ_{top} on top surface	0	0.27
ρ_{side} on side wall	0.5 (with $\cos^2$)	0.57
ρ_{bottom} on bottom surface	1	1

Tab. 6.2.: Wall loading on the several walls of the impedance corrugation for SIM and SIE normalized to the bottom surface ρ_{bottom}

6.2.2. Steady state temperature distribution

The cooling liquid, usually deionized water, is assumed to flow contrary within two concentric channels along the coaxial insert. The layout is assumed to be similar to the rod employed in the European 2 MW 170 GHz ITER design [PDD$^+$04], and the corresponding geometrical parameters can be found for the unrolled cross-section in Fig. 6.12. The radial width of the outer cooling channel is given as l_{W} and the wall thickness towards the impedance corrugation as l_{G}. The influence and heat exchange towards the inner (backward) water channel is neglected and set to be isothermal in order to simplify the calculation model. The turbulence of the water inside the outer cooling channel determines the heat exchange coefficient h_c and therefore the efficiency of the cooling. The turbulence described by the Reynold's number Re is approximately only dependent on the thickness l_{W}

of the channel and not on its absolute radii. The temperature along the cooling channel is assumed to be uniform and the temperature gradient in the cooling liquid is supposed to be small. The assumption of a constant water temperature is reasonable because the relative time when the water is close to the location of high field amplitude in the cavity is relatively short. If the flow rate of the water is too high, the insert starts to vibrate. A conservative value for the flow rate is chosen to be $40\,\mathrm{L/min}$, which corresponds to a velocity $v_\mathrm{W} = 5.9\,\mathrm{m/s}$ of the water inside the outer cooling channel. The specific deformations on the several walls of the impedance corrugation can be given as Δl towards the several directions. Δl_R therefore describes the deformation at the top wall towards R and Δl_φ the expansion at the side wall towards φ.

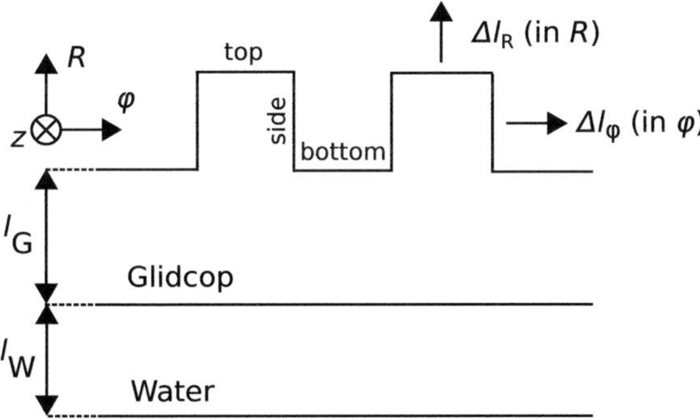

Fig. 6.12.: Unrolled cross-section of the coaxial insert (corrugation strongly magnified)

In a first step, the Reynold's number Re and the heat exchange coefficient h_c on the outer material/water interface are calculated [uC06]. Tab. 6.3 shows the dependency on the water flow rate for a fixed thickness of the cooling channel of $l_\mathrm{W} = 2.0\,\mathrm{mm}$. The turbulence and corresponding heat exchange rise strongly with higher water velocity v_W. Turbulent conditions within the cooling liquid can be assumed, because $Re > 10^4$ is determined for all considered flow velocities. Tab. 6.4 shows the influence of the thickness of the cooling channel at a fixed velocity of the water flow. For all approaches, the outer radius of the coaxial insert is set to be 13.0 mm as described in Sec. 2.5 and $l_G = 3.0$ mm. Due to the relation of the channel's outer and inner diameter, Re increases for an increasing water channel thickness l_W, but the heat exchange coefficient decreases. Within an experimental setup, this effect can be easily compensated due to an increase of the water velocity v_W.

v_W [m/s]	Flow rate [L/min]	Re	h_c [W/$^\circ$Cmm^2]
4.00	27.1	15505	0.0166
5.00	33.9	19382	0.0200
5.50	37.3	21321	0.0216
5.90	40.0	22871	0.0229
6.50	44.1	25197	0.0249
7.00	47.5	27135	0.0264

Tab. 6.3.: Re and h_c depending on the velocity of the cooling water inside the coaxial insert for a channel width of $l_W = 2.0$ mm

Fig. 6.13 displays the maximum surface temperature along the coaxial insert for the different models with a fixed heat exchange coefficient $h_c = 0.229$ W/$^\circ$Cmm^2 towards the cooling channel, which corresponds to a water velocity of $v_W = 5.9$ m/s. For the same loading on the bot-

l_W [mm]	Flow rate [L/min]	Re	h_c [W/$^\circ$Cmm^2]
1.60	32.7	18297	0.0237
1.80	36.4	20584	0.0233
2.00	40.0	22871	0.0229
2.20	43.6	25159	0.0226
2.40	47.0	27446	0.0223

Tab. 6.4.: Re and h_c depending on radial width l_W of the water channel for $v_W = 5.9$ m/s

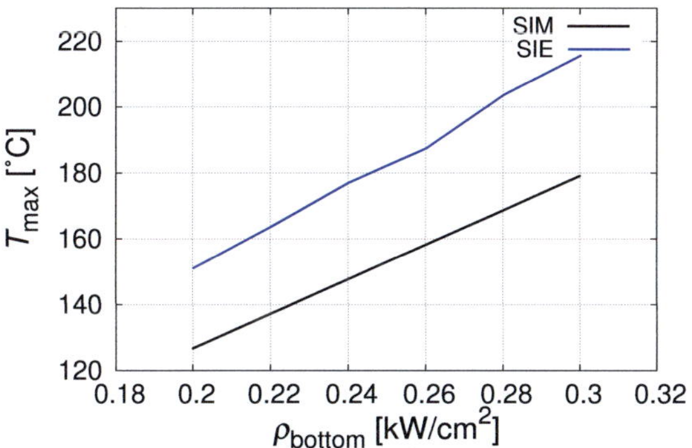

Fig. 6.13.: T_{max} on coaxial insert for different models and distribution of wall loading according to Tab. 6.2 ($h_{\mathrm{c}} = 0.229\,\mathrm{W}/\mathrm{°Cmm}^2$ for both)

tom surface of the grooves the models predict different distribution and a different summation of the loading on the side and top surfaces (Tab. 6.2). The ratios of the heat fluxes on the different walls normalized to the highest value on the bottom surface have to be chosen accordingly.

The results for the following calculations are always displayed in two diagrams for the SIM model on the left (a) and for the SIE model on the right (b). The loading on the several walls of the grooves are adjusted for every model in order to assume equal overall heating. The overall average heating over one period of the impedance corrugation assuming equal length of the four walls ($l = d = s/2$) is given as:

$$\bar{p} = \frac{p_{\mathrm{bottom}} + p_{\mathrm{top}} + 2p_{\mathrm{side}}}{4} \tag{6.17}$$

In addition, the calculations are performed for two different thicknesses l_{G} of the first material wall. In all calculations the initial temperature T_0 is set to room temperature equal $21.0\,\mathrm{°C}$.

Fig. 6.14(a) and 6.14(b) show the maximal temperature T_{max} on the surface of the insert over the velocity v_W of the cooling water. Each v_W-value corresponds to a flow rate, a Reynold's number Re and consequently to a specific heat exchange coefficient h_c. For lower flow rates, the maximum surface temperature rises strongly. Thus, v_W has to be checked carefully during long-pulse operation. If the width of the outer material is increased from 2.0 mm to 3.0 mm, the corresponding maximum surface temperature rises approximately by 5 °C.

The maximal temperature T_{max} on the surface of the insert depending on the width l_W of the cooling channel is displayed in Fig. 6.15(a) and 6.15(b). The turbulence of the cooling water, and therefore the heat exchange rate h_c, depends on the width of the ring-formed channel and not on its absolute radii. The width of the cooling channel has only small influences on the heat exchange coefficient and the resulting surface temperature. The geometry of the insert has to be designed in order to assure sufficient mechanical strength. The necessary heat exchange coefficient can be easily adjusted by the velocity of the cooling water.

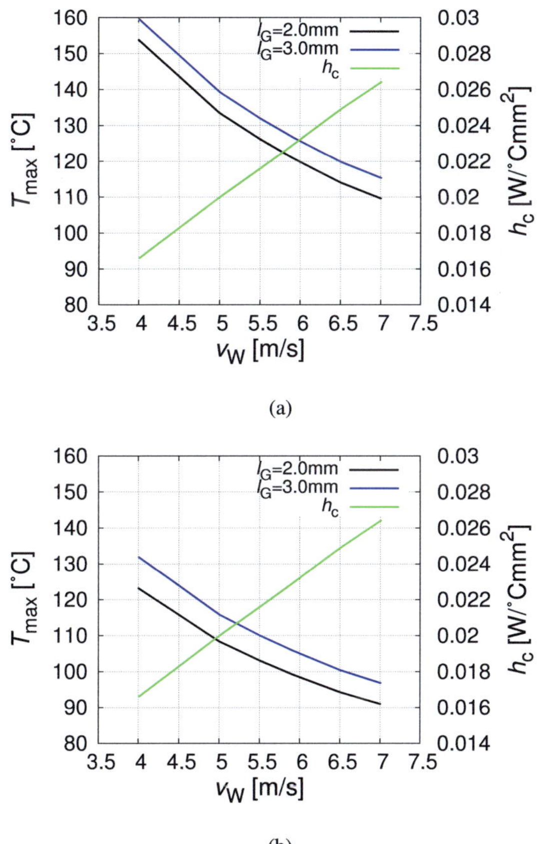

(a)

(b)

Fig. 6.14.: T_{max} on coaxial insert versus cooling water velocity v_{W}
(a: SIM with $\rho_{\mathrm{bottom}} = 0.2\,\mathrm{kW/cm^2}$; b: SIE with
$\rho_{\mathrm{bottom}} = 0.16\,\mathrm{kW/cm^2}$)

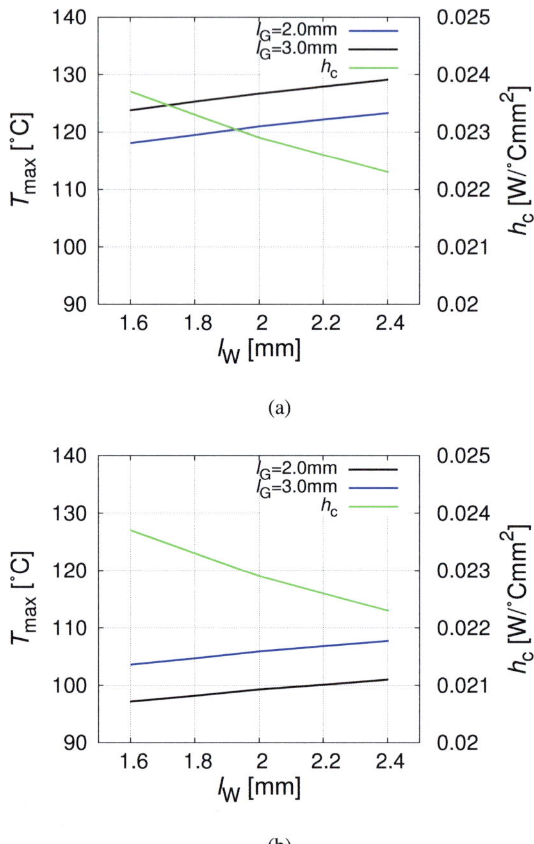

(a)

(b)

Fig. 6.15.: T_{max} on coaxial insert versus width of cooling channel l_W for $v_W = 5.9\,\mathrm{m/s}$ (a: SIM with $\rho_{bottom} = 0.2\,\mathrm{kW/cm^2}$; b: SIE with $\rho_{bottom} = 0.16\,\mathrm{kW/cm^2}$)

6.2.3. Deformation of the impedance corrugation

The fine grooves forming the corrugation are deformed, when the coaxial insert is heated during long pulse operation. The deformations, as displayed in Fig. 6.16(a) and 6.16(b), are split into two portions: $\Delta l_{\text{top,R}}$, which describes the deformation on the top surface of the fins in R-direction, and $\Delta l_{\text{side},\varphi}$, which describes the thermal expansion of the side surface in φ-direction. Again, fins with quadratic cross-section ($l = d = s/2$) and an edge length of $0.39\,\text{mm}$ are considered. The nominal cooling parameters are $h_{\text{c}} = 0.0229\,\text{W}/^\circ\text{Cmm}^2$ corresponding to $v_{\text{W}} = 5.90\,\text{m/s}$. Thus, the relative deformations of the impedance corrugation will be 3.1% in R- and 1.6‰ in φ-direction. The size of the expected thermal deformation will be small in comparison to wavelength and the original corrugation geometry. The same overall loading of the insert is assumed for both models ($\rho_{\text{SIM,bottom}} = 0.2\,\text{kW/cm}^2$ and $\rho_{\text{SIE,bottom}} = 0.16\,\text{kW/cm}^2$). A reduced efficiency of the corrugation on the eigenvalues of the modes due to thermal deformation is therefore not expected. In contrast, the wall loading and the corresponding surface temperature rises strongly with an increasing outer radius of the coaxial insert. A misalignment of the rod and a radius, which exceeds a safe value ($< 0.2\,\text{kW/cm}^2$) have to be carefully avoided.

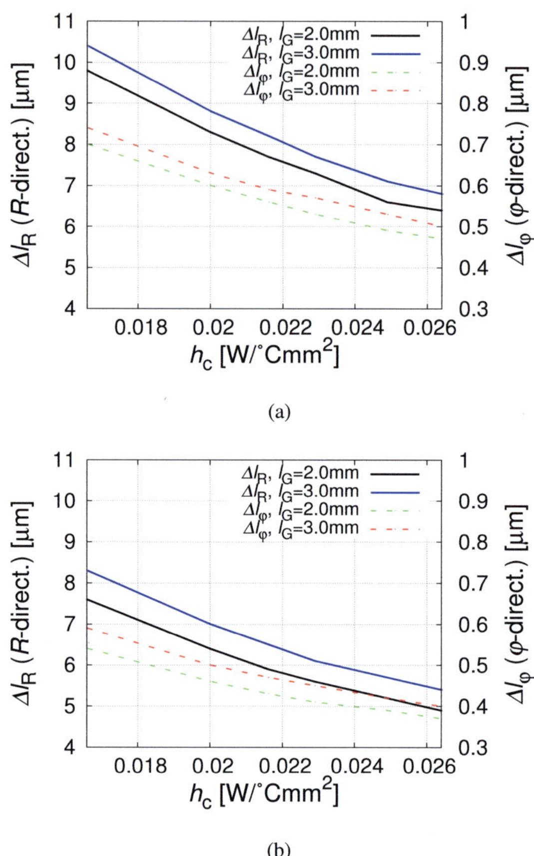

(a)

(b)

Fig. 6.16.: Deformation Δl on top towards R and on side towards φ (a: SIM with $\rho_{\mathrm{bottom}} = 0.2\,\mathrm{kW/cm^2}$; b: SIE with $\rho_{\mathrm{bottom}} = 0.16\,\mathrm{kW/cm^2}$)

6.3. Thermal distortion of the launcher surface

In the quasi-optical output system the high-order volume mode $TE_{-52,31}$ is transformed into two Gaussian-like beams. The transformation is achieved by small wall perturbations on the launcher surface which create a mode mixture as described in Ch. 4. At the entrance of the launcher the wall loading of the surface is uniform. The launcher surface is cooled by several water channels which surround the launcher along helical paths. Towards the launcher exit the beams get more and more focused and areas with high peak wall loading occur. This issue becomes especially critical in modern quasi-optical systems with highly focused beams. At the output section of a two-beam launcher for a $4\,\text{MW}$ gyrotron, relatively small sections on its surface reflect beams with $2\,\text{MW}$ of transmitted power each. In the following section the thermal deformations on the launcher layout from Sec. 4.3 are characterized and the influence of thermal expansions on the beam quality is determined.

Starting from the undeformed wall, as shown in Fig. 4.2, the field inside the launcher and the corresponding wall loading is calculated. The cooling channels are implemented as a double helix structure and are aligned between cylindrical ducts forming the inner and outer launcher surface. The channels are assumed to have a rectangular cross-section of $37.4 \times 5\,\text{mm}^2$ according to the size of the Brillouin zones. The heat exchange coefficient along the channels is calculated using analytic approximations [uC06], the corresponding heat exchange coefficients can be found in Tab. 6.5. Depending on the shape of the double helix structure, the bulks between two cooling channels can be beside (Fig. 6.17(a)) or centered (Fig. 6.17(b)) to the focus points. The alignment of the cooling channels is not suitable for real water flow but representative for possible geometries. The wall thickness of the launcher is assumed to be constant at $5.0\,\text{mm}$.

After calculation of the steady state temperature distribution on the launcher wall, the corresponding deformations are determined. In a next step, the new field profiles and the changes in the vector correlation coefficients using Eq. (4.42) for both beams are determined. The correlation between

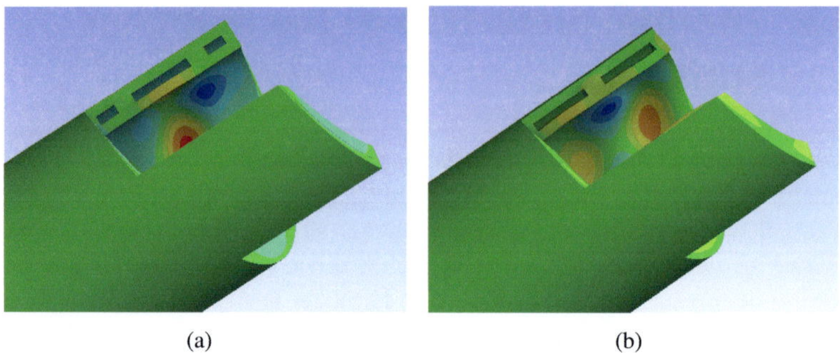

(a)   (b)

Fig. 6.17.: Channels at deformed launcher: bulks beside (a) and congruent (b) to focus points

deformed surface, corresponding new field distribution and wall loading is not solved self-consistent in order to simplify the problem and to achieve reasonable computation time.

Fig. 6.18(a) and 6.18(b) show the maximum surface temperature T_{max} on the launcher wall depending on the water velocity v_{W} in the cooling channels. The surface temperature inside the launcher rises strongly with less efficient cooling and the quality of both beams is significantly reduced. Already small deviations in the heat exchange coefficient at the launcher's cooling result in reduced vector correlation coefficients of the beams relative to an ideal Gaussian distribution. If the bulks between the cooling channels are centered to the focus points, the resulting surface temperature is higher in comparison to an alignment beside, but the beam quality remains at more favorable values. The additional bulk material directly at the focus points leads to a beneficial stabilization with respect to the Gaussian content. Fig. 6.19(a) shows the field distribution (H_{z}) along the launcher wall for a strongly non efficient cooling ($h_{\mathrm{c}} = 0.0155\,\mathrm{W/^\circ Cmm^2}$) for bulks aligned beside the focus points. In this worst case scenario the vector correlation coefficients of both output beams are reduced to $\eta_1 = 88.2\%$ and $\eta_2 = 87.3\%$. As a consequence, the irradiated beams show a strong defocusing in comparison to the original surface in Fig. 6.19(b). The in-

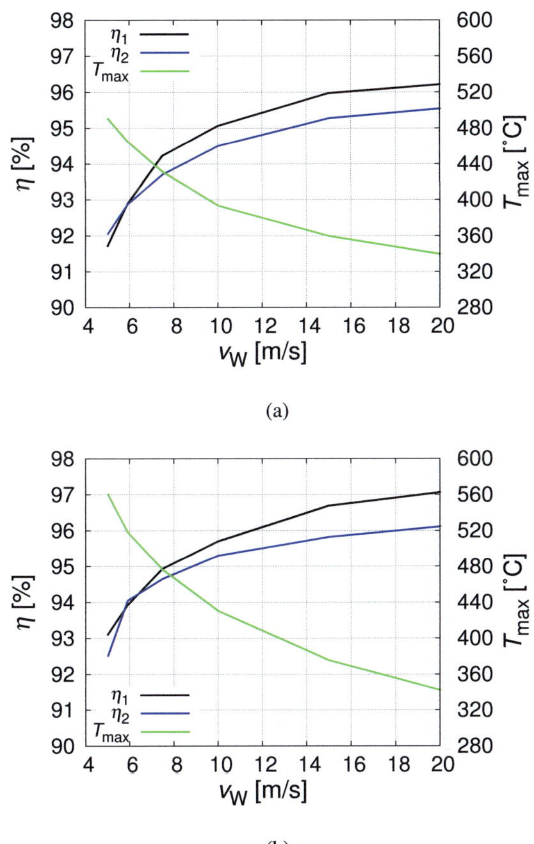

Fig. 6.18.: η for original launcher surface depending on v_W (a: bulks beside; b: bulks centered)

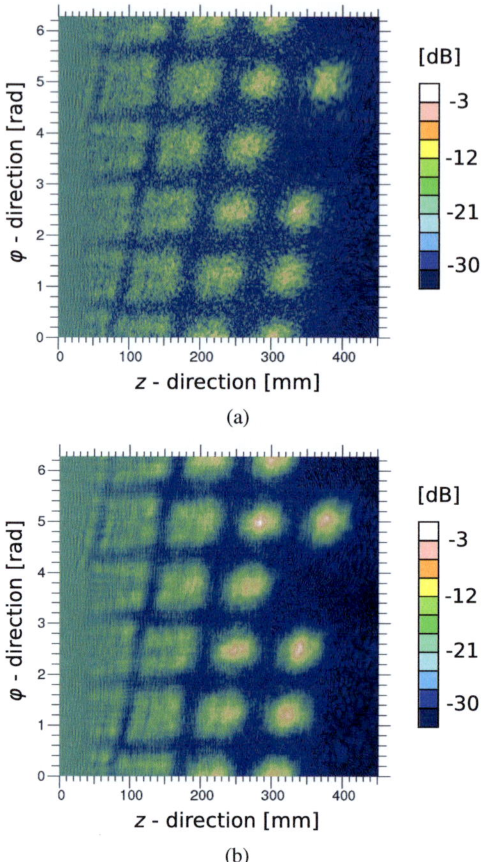

Fig. 6.19.: Magnetic field component H_z on strongly deformed launcher wall with $h_c = 0.0155\,\mathrm{W}/^\circ\mathrm{Cmm}^2$. (a) original surface with $\eta_1 = 88.2\%$ and $\eta_2 = 87.3\%$; (b) filtered with $\eta_1 = 94.0\%$ and $\eta_2 = 93.4\%$

v_W [m/s]	Flow rate [L/min]	h_c [W/$^\circ$Cmm^2]
5.00	56.0	0.0176
5.90	66.2	0.0201
7.50	84.1	0.0244
10.0	112.2	0.0307
15.0	168.2	0.0424
20.0	224.3	0.0534

Tab. 6.5.: h_c within the rectangular cooling channel (37.4×5 mm^2) for different cooling water velocities v_W

fluence of thermal-deformations on the launcher perturbations is therefore not negligible.

In a next step, a smoothened wall surface using the two-dimensional filter techniques, as introduced in Sec. 4.4.2, is considered and the corresponding thermal treatment is calculated. As shown before, the low-pass filtered surface structure delivers a high-quality beam even if the filter characteristic is chosen to be relatively small (here: $\omega_s \approx 0.4 \cdot \omega_n$). Fig. 6.20 shows the surface temperature and the vector correlation coefficients for both beams assuming the bulks between the cooling channels to be aligned beside the focus points. Fig. 6.20 is therefore directly comparable with Fig. 6.18(a) of the original optimized surface. The filtered surface results in nearly the same surface temperature for a changing cooling water flow, but the beam quality is up to a certain limit approximately independent of thermal deformation. The corresponding lower sensitivity to thermal characteristics make the utilization of a launcher with a filtered surface favorable. Fig. 6.19(b) displays the worst-cast scenario again with $h_c = 0.0155$ W/$^\circ$Cmm^2 for the deformed filtered surface. Even after strong thermal deformation the beams have a relatively high Gaussian content of $\eta_1 = 94.0\%$ and $\eta_2 = 93.4\%$. The filtered field profile shows significantly lower noise in the outer regions of the Brillouin zones where the filtering is mostly effective. In summary, the efficiency of the mode conversion is strongly dependent on the applied launcher cooling techniques.

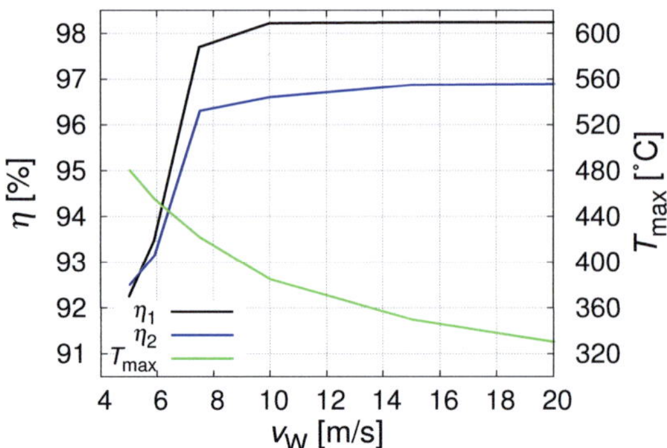

Fig. 6.20.: η for filtered launcher surface depending on v_W (bulk beside focus points)

Already small deviations from the necessary heat exchange coefficient reduce the beam quality significantly. The systematic reduction of steep gradients and deep grooves along the surface strongly reduces the launcher's sensitivity to thermal distortions.

7. Conclusion and outlook

In this work the feasibility of a $4\,\mathrm{MW}$ $170\,\mathrm{GHz}$ coaxial-cavity gyrotron for continuous wave operation is demonstrated as one step towards reliable and efficient heating sources, as they gain increasing relevance for the successful realization of future thermo-nuclear fusion reactors. For the first time complete physical designs of the major gyrotron components are elaborated considering long-pulse operation effects.

In a first step, one possible new operating mode is determined, followed by the development of detailed physical designs of the major gyrotron components: Electron gun, coaxial cavity, quasi-optical output coupler and collector. In addition, several thermo-mechanical studies are performed to identify challenges regarding cooling techniques and long-pulse operating effects.

A comprehensive mode selection procedure, based on a formalism of normalized variables, is utilized. It suggests one mode, namely the $TE_{-52,31}$, which is sufficiently separated from its competitors and supports an advanced two-beam quasi-optical mode converter. Due to its spread angle of nearly $144°$ between two reflection points on the launcher's surface in the quasi-optical ray representation, the $TE_{-52,31}$ mode is well suited for the conversion into two linearly polarized fundamental Gaussian output beams. By utilization of time-dependent and self-consistent approaches, a coaxial cavity is optimized and stable single-mode operation at $4.35\,\mathrm{MW}$ of generated output power with an interaction efficiency of 33% is achieved. A peak wall loading for realistic dispersion strengthened copper of less than $2.0\,\mathrm{kW/cm^2}$ for the cavity surface and less than $0.2\,\mathrm{kW/cm^2}$ at the coaxial insert, both relevant for CW operation, is realized. During the cavity optimization process it is especially critical to suppress undesired backward oscillations in the downtaper section. Nevertheless, stable and efficient single-mode operation is possible for a sufficiently wide range of operating parameters. The nonlinear cavity uptaper is optimized to guarantee low mode conversion towards the quasi-optical output coupler and a mode purity of 99.7% is attained at the launcher input. The RF-beams can be extracted from the gyrotron horizontally through two edge-cooled single-disk CVD-diamond windows. Due to the fixed thickness of the output windows tuned for the main frequency of 170 GHz, additional transmission at 136 GHz and 204 GHz is possible. The operating modes $TE_{-42,25}$ and $TE_{-62,37}$, which have approximately the same relative caustic radius $(m/\chi_{m,p})$ in the cavity and a similar spread angle for their conversion into fundamental Gaussian beams, are identified to be well suited for these supplementary operating frequencies. With an additional corrugation at the cavity's outer wall it is possible to further broaden the mode's stability area.

To provide the necessary hollow electron beam, designs for diode- and triode-type magnetron injection guns are elaborated using newly implemented optimization algorithms. Both electron guns deliver high-quality beams with significantly low transverse velocity spread of $\Delta\beta_{\perp,\mathrm{rms}} < 1.2\%$. At the necessary high electric field values of the 4 MW operating point, the triode-type layout shows a novel possibility for the reduction of its space requirements of approximately 30% in comparison to a diode-type design. The triode's additional modulation anode can be used to compensate the high electric field value in the vicinity of the emitter. The possible size reduction permits the utilization of a smaller warm bore hole of the superconducting magnet and thus reduces its overall costs. For the design of complete heating systems for future fusion reactor this can lead to significant cost savings.

After efficient electron-cyclotron-interaction in the coaxial cavity, the RF-field is guided by an optimized nonlinear uptaper and converted into two Gaussian beams by a highly efficient quasi-optical mode converter consisting of an advanced waveguide launcher. The launcher's surface perturbation is optimized on the basis of the scalar diffraction equation considering an oversize factor of 1.07. The untapered launcher has a length of 392 mm and two helical cuts at its end. The two beams have a vector correlation coefficient with respect to the fundamental Gaussian distribution of 98.3% and 96.9% respectively. Two-dimensional discrete low-pass filter functions are introduced to smoothen the launcher's surface perturbations after optimization. It is thereby possible to reduce steep gradients and deep grooves along the surface without decreasing the launchers overall efficiency.

Two collector layouts are optimized employing normal and dispersion strengthened copper as the absorption material. The designs are optimized using longitudinal magnetic sweeping systems with wobbled coil currents and optimized collector surface shapes. The admissible limit for the overall wall loading in the collector is only achievable with a high depression voltage which requires in turn a high quality of the electron beam. The copper collector layout with an average wall loading $< 500\,\mathrm{W/cm^2}$ has an inner radius of $400\,\mathrm{mm}$ and an absorbing length of approximately $1.0\,\mathrm{m}$ along the gyrotron axis. The higher admissible limit for dispersion strengthened copper of $< 1000\,\mathrm{W/cm^2}$ allows a more compact collector design with $300\,\mathrm{mm}$ in radius and $0.8\,\mathrm{m}$ in length.

For the characterization of several long-pulse operation effects, various computing scripts are introduced to enable the data exchange between commercial finite-element approaches and corresponding KIT in-house codes. The expected surface temperature of the optimized coaxial $4\,\mathrm{MW}$ cavity is strongly sensitive to its surface roughness and the efficiency of the applied cooling technique. The frequency shift due to thermal expansion of the cavity at nominal operating conditions with a peak wall loading of $1.7\,\mathrm{kW/cm^2}$ for realistic dispersion strengthened copper is determined to be approximately $-130\,\mathrm{MHz}$. In addition, the resonator shows a strongly increasing quality factor for a reduced heat exchange coefficient at the cylindrical cavity section. In order to guarantee stable operation, the efficiency of the cavity cooling and the inner wall's surface roughness have to be controlled carefully. The utilized calculation techniques are verified with experimental data available for the $1\,\mathrm{MW}$ $140\,\mathrm{GHz}$ series tube #1 for the stellarator Wendelstein 7-X and show satisfactory correlation.

During CW gyrotron operation, the coaxial insert and its impedance corrugation for advanced mode competition are heated and deformed. Based on two different models for the loading on the insert, the surface temperature is calculated and shows a strong dependence on the velocity of the cooling liquid in the insert's cooling channel. At nominal parameters the surface temperature reaches $120\,^{\circ}\mathrm{C}$ and the corresponding deformation of the several grooves of the impedance corrugation is not critical.

The optimized inner surface structure of the quasi-optical output launcher consists of fine perturbations in the scale of several tens of millimeter. Towards the end of the launcher, both beams are highly focused resulting in high local wall loading and a corresponding deformation of the optimized profile. The cooling channels with a rectangular cross-section are aligned as a double helix structure around the output coupler. The surface temperature at the launcher's focus points reaches $410\,°C$ for nominal operating conditions. The vector correlation coefficient in relation to the ideal Gaussian distribution and therefore the quality of both beams decreases strongly with reduced efficiency of the cooling structure. As a consequence, adequate launcher cooling is necessary in order to guarantee efficient conversion of the high-order cavity mode into two Gaussian beams and to minimize the generated stray radiation. After application of the newly developed two-dimensional filter techniques, the simplified launcher surface can operate more reliably at higher temperatures and corresponding thermal deformations.

During the next development phase, advanced shaping of the cavity's downtaper section might be introduced in order to further suppress unwanted backward waves. Even if stable and efficient operation is possible with the optimized cavity geometry, additional detailed investigations are desirable. A highly closed downtaper or a possible iris shaped layout might be reasonable approaches. RF-oscillations after the cavity (ACI) have to be considered and avoided by utilization of a realistic tapered magnetic field towards the launcher section in multi-mode self-consistent calculations. To further increase the fundamental Gaussian mode content of every output beam and to reduce the stray radiation, the layout of a suitable mirror system needs to be optimized. Within future steps, technical designs have to be achieved in order to transcribe the physical designs and the corresponding assumptions and geometrical simplifications into more concrete approaches. Future fusion reactors might also demand heating sources at higher frequencies (e.g. $200\,GHz$), and the presented designs would have to be adapted and verified consequently.

List of Figures

List of Tables

Bibliography

[AS72] M. Abramowitz and I.A. Stegun. *Handbook of Mathematical Functions with Formulas, Graphs, and Mathematical Tables.* Dover Publications, New York, USA, 1972.

[Bal38] C.A. Balanis. *Advanced Engineering Electromagnetics.* Hoboken, NJ, USA, 1938.

[BD04] A.A. Bogdashov and G.G. Denisov. Asymptotic Theory of High-Efficiency Converters of Higher-Order Waveguide Modes into Eigenwaves of Open Mirror Lines. *Radiophysics and Quantum Electronics*, Vol. 47, No. 4:283–296, 2004.

[BHM$^+$09] M. Brueck, T. Hilble, S. Morel, J.-M. Roquais, and H. Seidel. Verification of the TED Cathode Models by Longterm Lifetest Data. *Proceedings of the 10th International Vacuum Electronics Conference*, Rome, Italy:527–528, 2009.

[BL07] D. Bariou and C. Lievin. Design report ITER 170 GHz Gyrotron Contract No. EFDA-03/960. *Internal Report, Thales Electron Devices*, RT5307, 2007.

[Bor91] E. Borie. *Review of Gyrotron Theory.* Kernforschungszentrum Karlsruhe, Karlsruhe, Germany, 1991. KfK Report 4898.

[CB93] R.A. Correa and J.J. Barroso. Space Charge Effects of Gyrotron Electron Beams in Coaxial Cavities. *International Journal of Electronics*, Vol. 74:131–136, 1993.

[CD84] K.R. Chu and D. Dialetis. Theory of Harmonic Gyrotron Oscillator with Slotted Resonant Structure. *International Journal of Infrared and Millimeter Waves*, Vol. 5:37–56, 1984.

[CDK$^+$06] A.V. Chirkov, G.G. Denisov, M.L. Kulygin, V.I. Malygin, S.A. Malygin, A.B. Pavelyev, and E. A. Soluyanova. Use of Huygens's Principle for Analysis and Synthesis of the Fields in Oversized Waveguides. *Radiophysics and Quantum Electronics*, Vol. 49, No. 5:344–353, 2006.

[Che95] F.F. Chen. *Introduction to Plasma Physics*. Springer, New York, USA, 1995.

[Clo01] G. Le Clorarec. Characterisation des anneaux de Raschig dans une structure cavite-diode validation du choix pour le gyrotron 140 GHz. *Internal Report, Thales Electron Devices*, 2001.

[DAAea02] G. Dammertz, S. Alberti, A. Arnold, and E. Borie et al. Development of a 140-GHz 1-MW Continuous Wave Gyrotron for the W7-X Stellarator. *IEEE Transactions on Plasma Science*, 30, No. 3:808–818, 2002.

[DLM$^+$08] G.G. Denisov, A.G. Litvak, V.E. Myasnikov, E.M. Tai, and V.E. Zapevalov. Development in Russia of High-Power Gyrotrons for Fusion. *Nuclear Fusion*, Vol. 48, No. 5:55–63, 2008.

[Dru02] O. Drumm. *Numerische Optimierung eines quasi-optischen Wellentypwandlers für ein frequenzdurchstimmbares Gyrotron*. Forschungszentrum Karlsruhe, Karlsruhe, Germany, 2002. Scientific Report 6754.

[DS90] G. Dueck and T. Scheuer. Threshold Accepting: A General Purpose Optimization Algorithm Appearing Superior to Simulated Annealing. *Journal of Computational Physics*, Vol. 90:161–175, 1990.

[DT85] B.G. Danly and R.J. Temkin. Generalized Nonlinear Harmonic Gyrotron Theory. *Physics of Fluids*, Vol. 29:561–567, 1985.

[DZ04] O. Dumbrajs and G.I. Zaginaylov. Ohmic Losses in Coaxial Gyrotron Cavities with Corrugated Insert. *IEEE Transactions on Plasma Science*, Vol. 32, No. 3:861–866, 2004.

[Edg99] C.J. Edgcombe. *Gyrotron Oscillators - Their Principles and Practice*. Taylor and Francis Ltd., Cambridge, MA, USA, 1999. 3rd Edition.

[EMR93] G. Engeln-Müllges and F. Reutter. *Numerik-Algorithmen mit ANSI C-Programmen*. Mannheim, Germany, 1993.

[Erc02] V. Erckmann. The 10 MW ECRH System for W7-X. *Proceedings of the 29th IEEE International Conference on Plasma Science*, Banff, Alberta, Canada:244, 2002.

[ESL+07] V. Erckmann, M. Schmid, H.P. Laqua, G. Dammertz, S. Illy, H. Braune, F. Hollmann, F. Noke, and F. Purps. Advanced Gyrotron Collector Sweeping with smooth Power Distribution. *Proceedings of the 4th IAEA Technical Meeting on ECRH Physics and Technology for ITER*, Vienna, Austria, 2007.

[FGPY77] V.A. Flyagin, A.V Gaponov, M.I. Petelin, and Y.K. Yulpatov. The Gyrotron. *IEEE Transactions on Microwave Theory and Techniques*, MTT-25:514–521, 1977.

[FN88] V.A. Flyagin and G.S. Nusinovich. Gyrotron Oscillators. *Proceedings of the IEEE*, Vol. 76:644–656, 1988.

[Gap59] A.V. Gaponov. Interaction Between Electron Fluxes and Electromagnetic Waves in Waveguides. *Izv. Vyssh. Uchebn. Zahved. Radiofizika*, Vol. 2:450–462, 1959.

[GDF+10] G. Gantenbein, G. Dammertz, J. Flamm, S. Illy, S. Kern, G. Latsas, B. Piosczyk, T. Rzesnicki, A. Samartsev, A. Schlaich, M. Thumm, and I. Tigelis. Experimental Investigations and Analysis of Parasitic RF Oscillations in High-Power Gyrotrons. *IEEE Transactions on Plasma Science*, Vol. 38, No.6:1168–1177, 2010.

[Gei91]   T. Geist. *Hochfrequenz-messtechnische Charakterisierung von Herstellungsprozessen und Werkstoffen für Resonatoren eines 140 GHz Gyrotrons.* University Karlsruhe, Germany, 1991. Dissertation.

[HB92]   T.S. Huang and L.T. Bruton. *Two-Dimensional Digital Signal Processing.* Berlin, Germany, 1992.

[HG77]   J.L. Hirshfield and V.L. Granatstein. The Electron Cyclotron Maser - An Historical Survey. *IEEE Transactions on Microwave Theory and Techniques,* MTT-25:522–527, 1977.

[Hoe94]   O. Hoechtl. *Numerische Analyse der Modenkonversion in koaxialen Wellenleiterkomponenten.* Kernforschungszentrum Karlsruhe, Karlsruhe, Germany, 1994. KfK Report 5298.

[Iat96]   C.T. Iatrou. Operating-Mode Selection and Design of Gyrotron Oscillators. *Internal Report Forschungszentrum Karlsruhe (FZK),* F120.0022.012/A, 1996.

[IBD⁺97]   C.T. Iatrou, O. Braz, G. Dammertz, S. Kern, M. Kuntze, B. Piosczyk, and M. Thumm. Design and Operation of a 165 GHz 1.5 MW Coaxial-Cavity Gyrotron with Axial RF Output. *IEEE Transactions on Plasma Science,* Vol. 25, No. 3:470–479, 1997.

[Ill97]   S. Illy. *Untersuchungen von Strahlinstabilitäten in der Kompressionszone von Gyrotron-Oszillatoren mit Hilfe der kinetischen Theorie und zeitabhängiger Particle-in-Cell Simulation.* Forschungszentrum Karlsruhe, Karlsruhe, Germany, 1997. Scientific Report 6037.

[Jr.87]   A.S. Gilmour Jr. *Microwave Tubes.* Norwood, MA, USA, 1987.

[JTP⁺09]   J. Jin, M. Thumm, B. Piosczyk, S. Kern, J. Flamm, and T. Rzesnicki. Novel Numerical Method of the Analysis and Synthesis of the Fields in Highly Oversized Waveguide Mode

Converters. *IEEE Transactions on Microwave Theory and Techniques*, Vol. 57, No. 7:1661–1668, 2009.

[KAB+02] K. Koppenburg, A. Arnold, E. Borie, G. Dammertz, O. Drumm, M.V. Kartikeyan, B. Piosczyk, M. Thumm, and X. Yang. Design of Multifrequency High Power Gyrotron at FZK. *Proceedings of the 27th International Conference on Infrared and Millimeter Waves*, Vol. 1:153–154, 2002.

[KBT04] M. Kartikayan, E. Borie, and M. Thumm. *Gyrotrons - High-Power Microwave and Millimeter Wave Technology*. Berlin and Heidelberg, Germany, 2004.

[KDK+00] K. Koppenburg, G. Dammertz, M. Kuntze, B. Piosczyk, and M. Thumm. Fast Frequency-Step-Tunable High-Power Gyrotron with Hybrid Magnet System. *IEEE Transactions on Electronic Devices*, Vol. 48, No. 1:101–107, 2000.

[KDST85] K.E. Kreischer, D.G. Danly, J.B. Schutkeker, and R.J. Temkin. The Design of Megawatt Gyrotrons. *IEEE Transactions on Plasma Science*, Vol. 13:364–373, 1985.

[Ker96] S. Kern. *Numerische Simulation der Gyrotron-Wechselwirkung in koaxialen Resonatoren*. Forschungszentrum Karlsruhe, Karlsruhe, Germany, 1996. Scientific Report 5837.

[KGV83] S. Kirkpatrick, C.D. Gelatt, and M.P. Vecchi. Optimization by Simulated Annealing. *Science*, Vol. 220:671–680, 1983.

[KK09] K.-D. Kammeyer and Kristian Kroschel. *Digitale Signalverarbeitung - Filterung und Spektralanalyse mit MATLAB-Übungen*. Wiesbaden, Germany, 2009. 7th Edition.

[KLTZ92] A.N. Kuftin, V.K. Lygin, S.E. Tsimring, and V.E. Zapevalov. Numerical Simulation and Experimental Study of

Magnetron-Injection Guns for Powerful Short-Wave Gyrotrons. *International Journal of Electronics*, Vol. 72, No. 5,6:1145–1151, 1992.

[LC95] S. Lidia and R. Carr. Faster Magnet Sorting with a Threshold Acceptance Algorithm. *Review of Scientific Instruments*, Vol. 66:1865–1867, 1995.

[Lim90] J.S. Lim. *Two-Dimensional Signal and Image Processing*. NJ, USA, 1990.

[LJR⁺10] G. Li, J. Jin, T. Rzesnicki, S. Kern, and Manfred Thumm. Analysis of a Quasi-Optical Launcher Toward a Step-Tunable 2-MW Coaxial Cavity Gyrotron. *IEEE Transactions on Plasma Science*, Vol. 38, No. 6:1361–1368, 2010.

[LL66] L.D. Landau and E.M. Lifschitz. *Lehrbuch der theoretischen Physik Vol. 2*. Verlag Harri Deutsch, Berlin, Germany, 1966.

[Lyg95] V.K. Lygin. Numerical Simulation of Intense Helical Electron Beam with Calculation of the Velocity Distribution Function. *International Journal of Infrared and Millimeter Waves*, Vol. 16:363–376, 1995.

[MC77] J. McClellan and D. Chan. A 2-D FIR Filter Structure Derived from the Chebyshev Recursion. *IEEE Transactions on Circuits and Systems*, CAS-24:372–384, 1977.

[MF53] P.M. Morse and H. Feshbach. *Methods of Theoretical Physics*. McGraw-Hill Book, New York, USA, 1953.

[Mic98] G. Michel. *Feldprofilanalyse und -synthese im Millimeterwellenbereich*. Forschungszentrum Karlsruhe, Karlsruhe, Germany, 1998. Scientific Report 6216.

[Nat04] United Nations. World Population to 2300. *Department of Economic and Social Affairs*, ST/ESA/SER.A/236, 2004.

[Nei04]   J. Neilson.   SURF3D and LOT: Computer Codes for Design and Analysis of High-Performance QO Launchers in Gyrotrons. *Proceedings of the Joint 29th International Infrared Millimmeter Wave Conference and 12th Int. Terahertz Electronic Conference*, Karlsruhe, Germany:667–668, 2004.

[Nei05]   J.M. Neilson. Electric Field Integral Equation Analysis and Advanced Optimization of Quasi-Optical Launchers used in High-Power Gyrotrons. *Quasi-Optical Control of Intense Microwave Transmission*, 44-63:55–63, 2005.

[NGT04]   G.S. Nusinovich, V. Granatstein, and R.J. Temkin. *Introduction to the Physics of Gyrotrons*. Baltimore, MD, USA, 2004.

[PAG+09]   O. Prinz, A. Arnold, G. Gantenbein, Y.-H. Liu, M. Thumm, and D. Wagner. 2009.

[PDD+04]   B. Piosczyk, G. Dammertz, O. Dumbrajs, O. Drumm, S. Illy, J. Jin, and M. Thumm.   A 2-MW 170-GHz Coaxial Cavity Gyrotron. *IEEE Transactions on Plasma Science*, Vol. 32, No. 3:413–417, 2004.

[PFK+01]   A.B Pavelyev, V.A. Flyagin, V.I. Khizhnyak, V.N. Manuilov, and V.E. Zapevalov. Investigations of Advanced Coaxial Gyrotrons at IAP RAS. *Proceedings of the 4th International Kharkov Symposium*, Vol. 5:507–512, 2001.

[PHA+10]   I.G. Pagonakis, J.-P. Hogge, S. Alberti, B. Piosczyk, S. Illy, S.Kern, and C. Lievin. An Additional Criterion for Gyrotron Gun Design. *Proceedings of the IEEE International Conference on Plasma Science*, Norfolk, VA, USA, 2010.

[PHG+09]   I.G. Pagonakis, J.-P. Hogge, T. Goodman, S. Alberti, B. Piosczyk, S. Illy, T. Rzesnicki, S. Kern, and C. Lievin. Gun Design Criteria for the Refurbishment of the First Prototype of the EU 170GHz/2MW/CW Coaxial Cavity Gyrotron for ITER. *Proceedings of the 34th International Conference on Infrared, Millimeter and Terahertz Waves*, Busan, Korea, 2009.

[Pie77] G. Piefke. *Feldtheorie*. Mannheim, Germany, 1977.

[Pie81] G. Piefke. *Zweitor Analyse mit Leistungswellen*. Stuttgart, Germany, 1981.

[Pio01] B. Piosczyk. A Novel 4.5 MW Electron Gun for a Coaxial-Cavity Gyrotron. *IEEE Transactions on Electron Devices*, Vol. 48, No. 12:293–294, 2001.

[Pio06] B. Piosczyk. A Coaxial Magnetron Injection Gun CMIG for a 2 MW 170 GHz Coaxial Cavity Gyrotron. *International Journal of Infrared and Millimeter Waves*, Vol. 27:1041–1061, 2006.

[PTVF92] W.H. Press, S.A. Teukolsky, W.T. Vetterling, and B.P. Flannery. *Numerical Recipes*. Cambridge, MA, USA, 1992.

[Roa82] G.F. Roach. *Green's Functions*. Cambridge, U.K., 1982. 2nd Edition.

[Rod09] J.C. Rode. *Design Study of Magnetron Injection Guns for a 4 MW 170 GHz Coaxial Gyrotron*. Forschungszentrum Karlsruhe, Karlsruhe, Germany, 2009. Diploma Thesis.

[Sch59] J. Schneider. Stimulated Emission of Radiation by Relativistic Electrons in a Magnetic Field. *Physical Review Letters*, Vol. 2:504–505, 1959.

[SID$^+$07] M. Schmid, S. Illy, G. Dammertz, V. Erckmann, and M. Thumm. Transverse Field Collector Sweep System for High Power Gyrotrons. *Fusion Engineering and Design*, Vol. 82, No. 5-14:744–750, 2007.

[Sil86] S. Silver. *Microwave Antenna Theory and Design*. Peter Peregrinus Ltd., London, UK, 1986.

[Siv65] D.V. Sivukin. *Review of Plasma Physics.* Leotowich, Springer, New York, USA, 1965.

[SK06] J.J. Schneider and S. Kirkpatrick. *Stochastic Optimization - Scientific Computing.* Berlin and Heidelberg, Germany, 2006.

[SKK⁺09] K. Sakamoto, A. Kasugai, K. Kajiwara, K. Takahashi, Y. Oda, K. Hayashi, and N. Konayashi. Progress of High Power 170 GHz Gyrotron in JAEA. *Nuclear Fusion*, Vol. 49:095019, 2009.

[SNJ10] O.V. Sinitsyn, G.S Nusinovich, and T.M. Antonsen Jr. Some Possibilities for Reducing After-Cavity Interaction in Gyrotrons. *Proceedings of the IEEE International Vacuum Electronics Conference*, Monterey, CA, USA:333–334, 2010.

[TAH⁺01] M. Thumm, A. Arnold, R. Heidinger, M. Rhode, R. Schwab, and R. Spoerl. Status Report on CVD-Diamond Window Development for High Power ECRH. *Fusion Engineering and Design*, 53:517–524, 2001.

[TCP97] M. Thumm, R.A. Cairns, and A.D.R. Phelps. Modes and Mode Conversion in Microwave Devices. *Generation and Application of High Power Microwaves*, Bristol, U.K.:121–172, 1997.

[Thu10] M. Thumm. *State-of-the-Art of High Power Gyro-Devices and Free Electron Masers, Update 2009.* Karlsruhe Institute of Technology, Karlsruhe, Germany, 2010. Scientific Report 7540.

[Twi58] R.Q. Twiss. Radiation Transfer and the Possibility of Negative Absorption in Radio Astronomy. *Australien Journal of Physics*, Vol. 11:567–579, 1958.

[TWMG94] T.M. Tran, D.R. Whaley, S. Merazzi, and R. Gruber. DAPHNE, a 2D Axissymmetric Electron Gun Simulation Code. *Proceedings of the 6th Joint EPS-APS In-*

ternational Conference on Physics Computing, Manno, Switzerland:491–494, 1994.

[TYA$^+$05] M. Thumm, X. Yang, A. Arnold, G. Dammertz, G. Michel, J. Pretterebner, and D. Wagner. A High-Efficiency Quasi-Optical Mode Converter for a 170 GHz 1 MW CW Gyrotron. *IEEE Transactions on Electron Devices*, Vol. 52, No. 5:818–824, 2005.

[uC06] VDI Gesellschaft Verfahrenstechnik und Chemieingenieur-wesen. *VDI Wärmeatlas*. Berlin, Germany, 2006. 10th Edition.

[VZO$^+$69] S.N. Vlasov, G.M. Zhislin, I.M. Orlova, M.I. Petelin, and G.G. Rogacheva. Irregular Waveguides as Open Resonators. *Radiophysics and Quantum Electronics*, Vol. 12:972–978, 1969.

[VZO76] S.N. Vlasov, L.I. Zagryadskaya, and I.M. Orlova. Open Coaxial Resonators for Gyrotrons. *Radio Engineering and Electronic Physics*, Vol. 21:96–102, 1976.

[Wei69] L.A. Weinstein. *Open Resonators and Open Waveguides*. The Golem Press, Boulder, CO, USA, 1969.

[Yee66] K.S. Yee. Numerical Solution of Initial Boundary Value Problems Involving Maxwell's Equations in Isotropic Media. *IEEE Transaction on Antennas and Propagation*, Vol. 14:302–307, 1966.

[Yee83] K.S. Yee. Finite-Difference Solution of Maxwell's Equations in Generalized Non-Orthogonal Coordinates. *IEEE Transactions on Nuclear Science*, Vol. 30, No. 6:4589–4591, 1983.

A. Appendix

A.1. Glidcop material properties

Glidcop is the registered trade name of SCM Metal Products for copper which is dispersion-strengthened with ultra-fine particles of aluminium oxide. Tab. A.1 lists material properties of different Glidcop types and includes the data for oxygen-free high thermal conductivity copper (OFHC). All data is given for room temperature 21 °C unless otherwise noted.

	OFHC	AL-15	AL-25	AL-60
Al_2O_3 content [wt.%]	0	0.7	1.2	2.7
Density [g/cm^3]	8.94	8.90	8.86	8.81
Melting Point [°C]	1083	–	1083	–
El. conductivity [MS/m]	58	54	50	45
Isotr. th. conductivity [W/°Cm]	391	365	344	322
Coeff. th. expansion [μm/m°C]	17.7	–	16.6	–
Young's Modulus [GPa]	115	–	130	–
Poissons's ratio (50 °C)	0.34	–	0.32	–
Specific heat [J/kg°C] (20 °C)	38.5	–	38.5	–
Ultimate tensile strength [MPa]	220	360	390	470

Tab. A.1.: Material properties for various grades of Glidcop

A.2. Geometrical parametrizations for the gun designs

Fig. A.1 and A.2 show the geometrical parametrizations of the cathode and the coaxial insert for the optimization process of the electron guns. These are common to both (diode- and triode-type) design approaches. The layout for the anode of the diode-type gun is shown in Fig. A.3. The geometry for the anode and modulation-anode for the triode-type MIG is given in Fig. A.4.

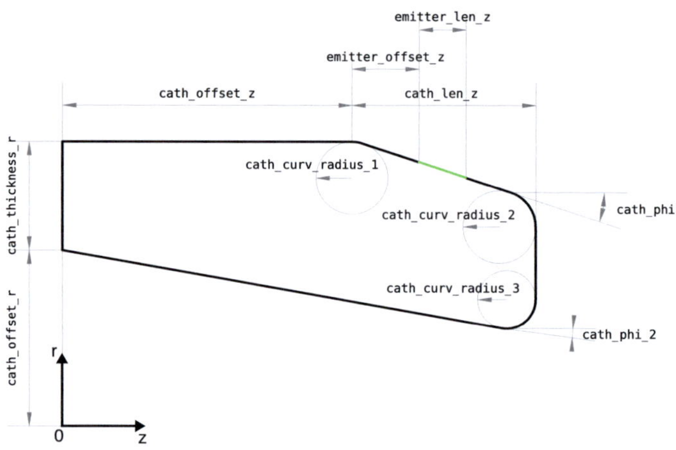

Fig. A.1.: Parameterization of the cathode for diode- and triode-type MIGs (not to scale)

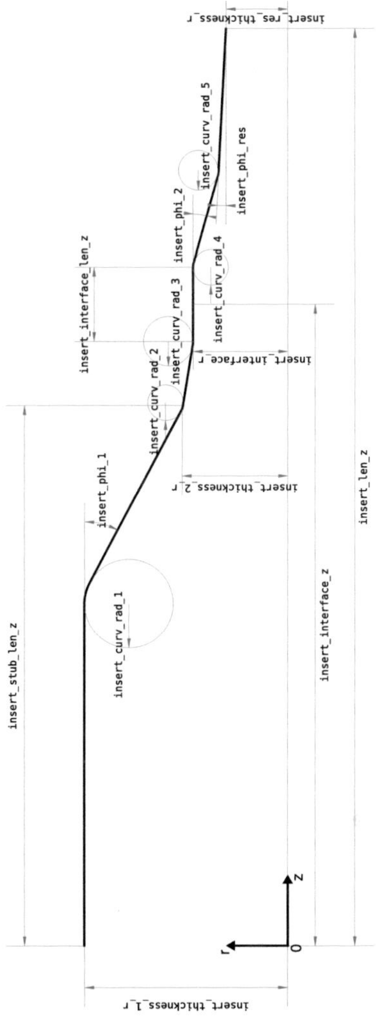

Fig. A.2.: Parameterization of the coaxial insert for diode- and triode-type MIGs (not to scale)

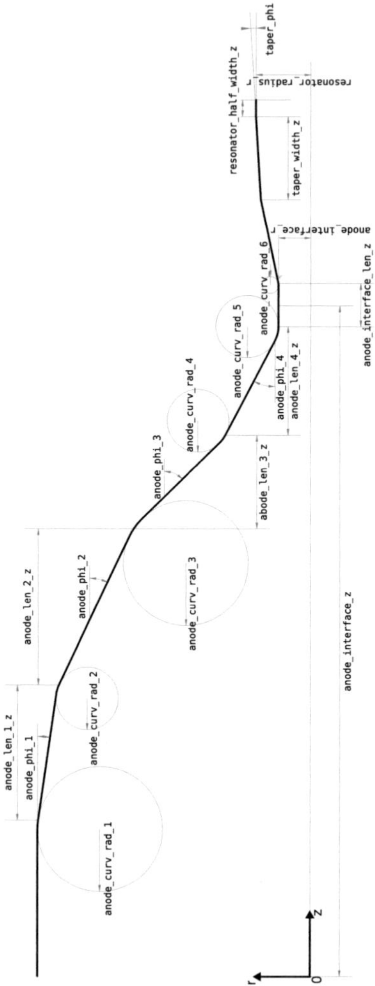

Fig. A.3.: Parameterization of the anode for diode-type MIG (not to scale)

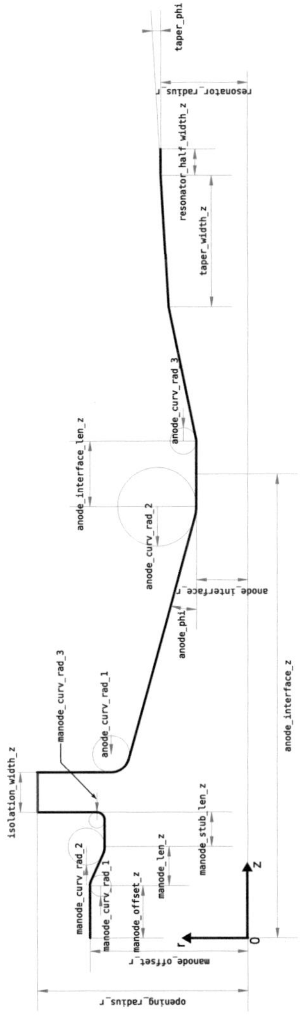

Fig. A.4.: Parameterization of the modulation anode and the anode for triode-type MIG (not to scale)

Acknowledgement

Die vorliegende Arbeit entstand im Rahmen des "European Fusion Training"-Programms der europäischen Kommission am Institut für Hochleistungsimpuls- und Mikrowellentechnik (IHM) des Karlsruher Instituts für Technologie (KIT).

Mein herzlicher Dank gilt Herrn Professor Dr. rer. nat. Dr. h.c. Manfred Thumm für die Betreuung dieser Arbeit und die Übernahme des Hauptreferats. Sein hervorragendes Engagement, sein großes Vertrauen und die Freiräume, die er mir bot, haben diese Arbeit erst möglich gemacht.

Herrn Professor Dr.-Ing. Mathias Noe, Leiter des Instituts für Technische Physik (ITEP) am Karlsruher Institut für Technologie (KIT), danke ich für die freundliche Übernahme des Korreferats.

Allen Mitgliedern der Gyrotronabteilung bin ich für das exzellente Arbeitsklima, die Kaffeerunden, die unterhaltsamen Dienstreisen und die große mir entgegengebrachte Hilfsbereitschaft dankbar. Ohne ihr Wissen und ihre Ideen wäre meine Arbeit nicht soweit gekommen. Herausheben möchte ich die unermüdliche Unterstützung von Dr. Stefan Kern, der immer wieder entscheidende Punkte beigetragen hat. Herrn Dr. Stefan Illy danke ich herzlich für die stetige Hilfe beim Kanonendesign. Jianbo and Jens, thanks for your help with this complicated round duct called launcher. Herrn Johann Christian Rode danke ich für die Beiträge während seiner Diplomarbeit.

I'm grateful to the team at Thales Electron Devices for all their help during my time in Paris. Thanks for teaching me all the fabrication techniques for high power microwave tubes as well as the French way of life. I owe my deepest gratitude to Christophe Liévin, for his support and confidence. Dem Karlsruhe House of Young Scientists (KHYS) danke ich hierbei für die zusätzliche finanzielle Unterstützung.

Unerlässlich für das Gelingen dieser Arbeit war der grenzenlose Rückhalt meiner Eltern, denen ich für Ihr Vertrauen und die fortwährende Unterstützung danken möchte. Erst ihre Aufopferung hat mich zu dem gemacht was ich heute bin. Ganz besonderen Dank an meinen Bruder Steffen, für all die Mühen und für jedwede Hilfe, die er jederzeit zu geben bereit ist.

Meinen Freunden danke ich für die Abwechslung und die langen Nächte, wodurch sie mich immer wieder erfolgreich abgelenkt und mich daran erinnert haben, was wirklich wichtig ist.

Schließlich danke ich von ganzem Herzen meiner Lebensgefährtin Kathrin, die mich mit all ihrer Kraft unterstützte und dadurch die Basis für diese Arbeit bildete. Ich bin stolz sie an meiner Seite zu wissen.